1982

A Combinatorial Introduction to Topology

A Series of Books in Mathematical Sciences
Victor Klee, Editor

A Combinatorial Introduction to Topology

Michael Henle

Oberlin College

W. H. Freeman and Company
San Francisco

Sponsoring Editor: Peter Renz
Project Editor: Betsy Dilernia
Copyeditor: Karen Judd
Designer: Laurence Hyman
Production Coordinator: M. Y. Mim
Illustration Coordinator: Batyah Janowski
Art Studio: Eric G. Hieber Associates, Inc.
Compositor: Syntax International
Printer and Binder: The Maple-Vail Book Manufacturing Group

Library of Congress Cataloging in Publication Data

Henle, Michael.
  A combinatorial introduction to topology.

  (A series of books in mathematical sciences)
  Bibliography: p.
  Includes index.
  1. Algebraic topology.   I. Title.
QA612.H46   514'.2   78-14874
ISBN 0-7167-0083-2

"Liberté, égalité, homologie."

Contents

Chapter Three

Plane Homology and the Jordan Curve Theorem

Chapter Four

Surfaces

Chapter Five

Homology of Complexes

Chapter Six

Continuous Transformations

Supplement

Topics in Point Set Topology

Preface

Topology is remarkable for its contributions to the popular culture of mathematics. Euler's formula for polyhedra, the four color theorem, the Möbius strip, the Klein bottle, and the general notion of a rubber sheet geometry are all part of the folklore of current mathematics. The student in a first course in topology, however, must often wonder where all the Klein bottles went, for such courses are most often devoted to **point set topology,** the branch of topology that lies at the foundation of modern analysis but whose intersection with the popular notions of topology is almost empty. In contrast, the present work offers an introduction to **combinatorial** or **algebraic topology,** the other great branch of the subject and the source of most of its popular aspects.

There are many good reasons for putting combinatorial topology on an equal footing with point set topology. One is the strong intuitive geometric appeal of combinatorial topology. Another is its wealth of applications, many of which result from connections with the theory of differential equations. Still another reason is its connection with abstract algebra via the theory of groups. Combinatorial topology is uniquely the subject where students of mathematics below graduate level can see the three major divisions of mathematics—analysis, geometry, and algebra—working together amicably on important problems.

In order to bring out these points, the subject matter of this introductory volume has been deliberately restricted. Most important is the emphasis on surfaces. The advantage is that the theorems can be easily visualized, encouraging geometric intuition. At the same time, this area is full of interesting applications arising from systems of differential equations. To bring out the interaction of geometry and algebra, it was decided to develop one algebraic tool in detail, homology, rather than several algebraic tools briefly. Thus an alternative title for the book could be *Homology of Surfaces*. These limitations of subject matter may be defended not only on the pedagogical grounds just outlined but also on historical grounds. Topological investigations of surfaces go back to Riemann and have always played an important role as a model for the study of more complicated objects, while homology, also one of the earliest areas of research, played a key role in the introduction of algebra into topology.

It is hoped that this book will prove useful to two classes of students: upper-level undergraduates as an introduction to topology, and first-year graduate students as a prelude to more abstract and technical texts. The prerequisites for reading this book are only some knowledge of differential equations and multivariate calculus. No previous acquaintance with topology or algebra is required. Point set topology and group theory are developed as they are needed. In addition, a supplement surveying point set topology is included both for the interested student and for the instructor who wishes to teach a mixture of point set and algebraic topology.

This book is suitable for a variety of courses. For undergraduates, a short course (8–10 weeks) can be based on the first four-and-a-half chapters (up to §28), while a longer course (12–15 weeks) could cover all the first five chapters as well as parts of Chapter Six (§§36, 37, and 38 are all logical stopping points). The Interdependency Chart (opposite) reveals a number of sections that can be omitted if time grows short, although inclusion of as many as possible of these application sections is recommended. The supplement can be used to replace four sections of algebraic topology if desired. It is independent of Chapters Two through Six and so can be taken up anytime after Chapter One. A graduate class can cover the book in less than a semester, especially if point set topology is assumed. The remainder of the course can be devoted to a more technical treatment of algebraic topology or to further investigation of the topics in this book. Suggestions for further reading are given at the end of each chapter as well as in special sections at the end of the book.

INTERDEPENDENCY CHART

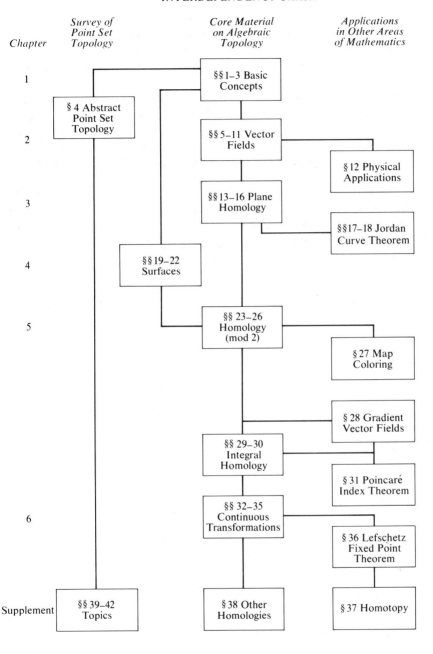

Chapter | Survey of Point Set Topology | Core Material on Algebraic Topology | Applications in Other Areas of Mathematics

1 §§ 1–3 Basic Concepts

§ 4 Abstract Point Set Topology

2 §§ 5–11 Vector Fields

§ 12 Physical Applications

3 §§ 13–16 Plane Homology

§§ 17–18 Jordan Curve Theorem

4 §§ 19–22 Surfaces

5 §§ 23–26 Homology (mod 2)

§ 27 Map Coloring

§ 28 Gradient Vector Fields

§§ 29–30 Integral Homology

§ 31 Poincaré Index Theorem

6 §§ 32–35 Continuous Transformations

§ 36 Lefschetz Fixed Point Theorem

Supplement §§ 39–42 Topics

§ 38 Other Homologies

§ 37 Homotopy

An important feature of the book is the problems. *They are an integral part of the book.* The argument of the text often depends upon statements made in the exercises. Therefore they should at least be read, if not worked out. A section of hints and answers (mostly hints) is provided at the end of the book.

The creation of algebraic topology is one of the triumphs of twentieth-century mathematics. The goal of this book is to show how geometric and algebraic ideas met and grew together into an important branch of mathematics in the recent past. At the same time, the attempt has been made to preserve some of the fun and adventure that is naturally part of a mathematical investigation.

I would like to thank Peter Renz and the rest of the staff at W. H. Freeman and Company for their encouragement and assistance. I particularly wish to thank Professors Victor Klee and Isaac Namioka for their careful reading of the manuscript. Finally, I wish to acknowledge the inspiration of Mrs. Helen Garstens, in whose eighth and ninth grade classes I first encountered topology.

Michael Henle
June 1977

A Combinatorial Introduction to Topology

one

Basic Concepts

§1 THE COMBINATORIAL METHOD

Topology is a branch of geometry. To get an idea of its subject matter, imagine a geometric figure, such as a circular disk, cut from a sheet of rubber and subjected to all sorts of twisting, pulling, and stretching. Any deformation of this sort is permitted, provided the rubber can withstand it without ripping or tearing. Figure 1.1 shows some of the shapes the disk might assume. The rubber must be very flexible! Perhaps silly putty would work. Technically these distortions are called continuous transformations. The formal definition of **topology** is the study of properties of figures that endure when the figures are subjected to continuous transformations. Technical details will come later (§2). In this section we wish to describe some of the ideas of topology at the intuitive level where topology has earned the nickname "rubber sheet geometry."

Figure 1.1   A disk and some topological equivalents.

From Figure 1.1 it is clear that a continuous transformation can destroy all the usual geometric properties of the disk, such as its shape, area, and perimeter. Therefore these are *not* topological properties of the disk. Any topological properties the disk has must be shared with the other shapes in Figure 1.1. These shapes are called topologically equivalent, meaning that any one can be continuously transformed to any other. Although topology will seem very different from Euclidean geometry, the two are fundamentally alike. In Euclidean geometry there is also a basic group of transformations, called congruences. The subject matter of Euclidean geometry is determined by the congruences, just as the subject matter of topology is determined by the continuous transformations. The congruences all preserve distances. For this reason they are also called rigid motions. Under a rigid motion a disk remains a circular disk with the same shape, radius, area, and perimeter. Therefore these are Euclidean properties of the disk. Euclidean geometry can be defined as the study of those properties of figures that endure when the figures are subjected to rigid motions.

What is a topological property of the disk? For one thing, the disk is in one piece. This is a topological property, since any attempt to divide the disk into more than one piece would require cutting or tearing, and these are forbidden in continuous transformations. Another topological property of the disk is that it has just one boundary curve. Another plane figure of some importance in topology, the **annulus** (ring shape) (see Figure 1.2), can be distinguished from the disk by this property, since the annulus has

Figure 1.2   An annulus and friend (another annulus).

two boundary curves. Another property distinguishing the disk from the annulus is that the latter divides the plane into two parts while the former does not.

Exercises

1. Examine some familiar objects topologically. For example, which of the following objects are topologically equivalent: hand iron, baseball bat, pretzel, telephone handset, rubber band, chair (consider several types), funnel, a scissor, Frisbee, etc.?

2. Why is it said that a topologist is someone who can't tell a doughnut from a coffee cup?

Let us call any figure topologically equivalent to a disk a **cell**. A cell is one of the simplest of topological figures. The first principle of combinatorial topology is to study the complicated figures that can be built in some way from simple figures. To put this principle into practice in this book, we will restrict ourselves to figures that can be constructed from cells by gluing and pasting them together along their edges. Figure 1.3 gives some examples. The edges of the cells sewn together to make each example are clearly marked. The numbers of cells used are 4, 7, 8, 8, 3, and 5, respectively. These numbers are not rigidly determined. For example, it would be simpler to build a sphere from just two cells, like the northern and southern hemispheres of a globe, or the two hourglass-shaped cells that are sewn together to make a baseball (Figure 1.4). Figures like these are called **complexes**. All but the most pathological of geometric surfaces are complexes; that is, they can be regarded as built in this way from cells.

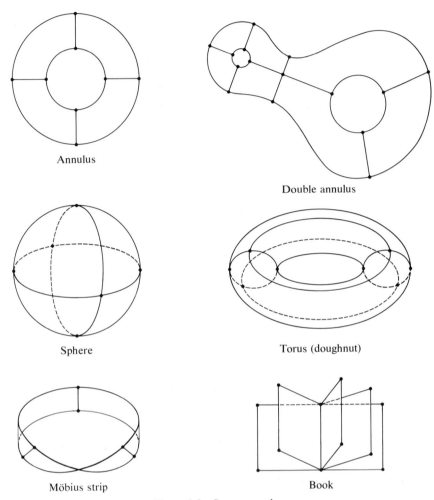

Annulus

Double annulus

Sphere

Torus (doughnut)

Möbius strip

Book

Figure 1.3 Some complexes.

Figure 1.4 Another sphere.

Exercises

3. Examine the surfaces presented by some familiar objects. In each case try to show how the surface can be constructed from cells.
4. Adopt the rule that no cell is to be sewn to itself. Find the minimum number of cells needed to make an annulus, a double annulus, and so forth. Do these numbers change if cells are permitted to be sewn to themselves?

The combinatorial method is used not only to construct complicated figures from simple ones but also to deduce properties of the complicated from the simple. This idea is actually common to much of mathematics. In combinatorial topology it is remarkable that the only machinery needed to make these deductions is the elementary process of counting! This is the miracle of combinatorial topology, that counting alone suffices to found a sophisticated geometric theory. This reliance on counting is what distinguishes combinatorial topology from other branches of topology and geometry.

In order to count we must have something to count. To this end we jazz up the cells a bit. We call a cell a **polygon** when a finite number of points on the boundary are chosen as vertexes. The sections of boundary in between vertexes are then called **edges**. A polygon is called an **n-gon**, where n is the number of vertexes. The complexes in Figure 1.3 are actually built from cells in the form of polygons. Thus the annulus there is composed of four 4-gons, while the double annulus is composed of three 4-gons and four 5-gons. As a further example, Figure 1.5 shows a set of four 3-gons, a 4-gon, and a 5-gon that have often been successfully sold commercially as a puzzle. The same figure shows two complexes (among many) that can be made from these polygons: a rectangle and a letter T. Of course, considered topologically these two complexes are equivalent. In fact, after the pieces of the puzzle are sewn together, both complexes are cells.

The choice of certain points as vertexes in order to be able to regard cells as polygons may seem irrelevant to topology. However, it provides something to count and so affords the combinatorial method something with which to work. In the future we assume that all complexes are formed from polygons, and furthermore we adopt the rule that vertexes are sewn to vertexes and whole edges are sewn to whole edges.

As an application of the combinatorial method, here is a derivation of a famous result: Euler's formula for polyhedra. A **polyhedron** is a complex that

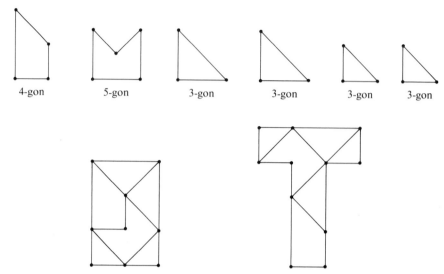

Figure 1.5   The T puzzle.

is topologically equivalent to a sphere. For example, the sphere in Figure 1.3 is a polyhedron, an **octahedron** in this terminology. Given a polyhedron, let F stand for the number of cells (called **faces** in this context), E the number of edges, and V the number of vertexes. Euler's formula states that

$$F - E + V = 2 \tag{1}$$

This remarkable result is the earliest discovery in combinatorial topology. It is due to Descartes (1639) but bears the name of Euler, who rediscovered it and published a proof (1751). The following proof is due to Cauchy (1811).

We begin by removing one of the faces of the polyhedron. The remainder is topologically equivalent to a cell and so may be flattened into a plane. For example, Figure 1.6 shows this operation being performed on a cube. The result is a cell in the plane divided into polygons. By removing one face and leaving edges and vertexes intact, the sum $F - E + V$ has been decreased by one. We must now prove in general for a complex equivalent to a cell that

$$F - E + V = 1 \tag{2}$$

This is called Euler's formula for a cell. It applies, for example, to the two complexes of Figure 1.5.

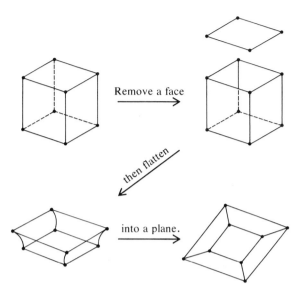

Figure 1.6 Proof of Euler's formula.
Step one: removing a face.

The next step is to triangulate the complex: divide each polygon into triangles by drawing diagonals, as in Figure 1.7. Each diagonal adds one edge and one face to the complex so that the quantity $F - E + V$ is unchanged by this process. In the final step, the triangles of the figure are removed one by one starting with those on the boundary. There are two types of removal, depending on whether the triangle being removed has one or two edges on the boundary (Figure 1.8). In the first case, the removal decreases both F and E by one, while in the second case, F and V decrease by one and E decreases by two. In any event, the quantity $F - E + V$ is

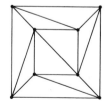

Figure 1.7 Step two:
the topless cube triangulated.

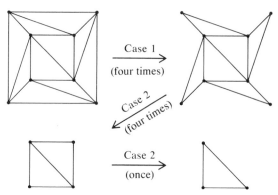

Figure 1.8   Step three: removal of triangles.

unchanged by the removal. Eventually we are left with just one triangle for which $F = 1$, $E = 3$, and $V = 3$. Obviously here $F - E + V = 1$. Since Steps 2 and 3 did not alter the sum $F - E + V$, this proves Euler's formula.

Exercises

5. Repeat Exercise 4, assuming this time that all cells are in the form of polygons and supposing that no cell can be sewn to itself or to any other cell along more than one edge.
6. Confirm Euler's formula for polyhedra in the cases of the cube and octahedron. Confirm Euler's formula for cells for the complexes in Figure 1.5.
7. Carry out the steps of the proof of Euler's formula for a tetrahedron (triangular pyramid) and an octahedron.

To appreciate the topological value of Euler's formula, consider it for a moment from the point of view of the sphere. Euler's formula says that no matter how the sphere may be divided into polygons, the sum of the number of faces minus the number of edges plus the number of vertexes is two: *always*. Rather like the surface area of the sphere, the number two is a property of the sphere itself, for although it is discovered only by dividing the sphere into polygons and then counting and adding faces, edges, and vertexes, the number two is independent of the manner of this division. Unlike surface area, however, the number two is also a *topological property of the sphere*, since the proof of Euler's formula is valid for any figure topologically equivalent to the sphere. Thus a topological property can result from the introduction of the seemingly irrelevant vertexes and edges. This is the whole point of the combinatorial method. Although its meaning is not completely clear yet, in fact *the number two is the single most important topological property of the sphere.* The remainder of this book may be regarded as an elaborate justification of this assertion. The number two is called the **Euler characteristic** of the sphere. Similarly, on account of (2), the Euler characteristic of the cell is one.

As an application of Euler's formula, we can determine all the regular polyhedra. A **regular** polyhedron is one in which all faces have the same number of edges, and the same number of faces meet at each vertex. The cube is an example of a regular polyhedron: all faces have four edges and all vertexes lie on three faces. The Greeks discovered five regular polyhedra. They are called the **Platonic solids**, and their construction is the climax of Euclid's Elements. We shall soon see why there are no others. Let a be the number of edges on each face of a regular polyhedron, and let b be the number of edges meeting at each vertex. Then the number aF counts all the edges face by face. Each edge is counted twice in this way, since each edge belongs to two faces. In other words, $aF = 2E$. Similarly, since each edge has two vertexes, $bV = 2E$. Euler's formula now reads

$$\frac{2E}{a} - E + \frac{2E}{b} = 2$$

or

$$\frac{1}{a} + \frac{1}{b} - \frac{1}{2} = \frac{1}{E} \tag{3}$$

This is an example of a **Diophantine equation**, so called because the solutions a, b, and E by the nature of the problem must be positive integers, and because equations with solutions of this type were studied by the Greek mathematician Diophantus. The restriction to the positive integers is an immense simplification. In this case it makes a complete solution possible. Actually both a and b must be greater than two, since faces or vertexes with only one edge are impossible, and those with just two edges are usually deliberately excluded. Since both a and b are greater than two, they must also be less than six in order that the left-hand side of (3) be positive. Thus when a and b are greater than two, there are only a finite number of possibilities, and the solutions may be found by trial and error. There turn out to be just five solutions, each corresponding to a polyhedron that can actually be constructed using Euclidean polygons. The solutions are illustrated in Figure 1.9.

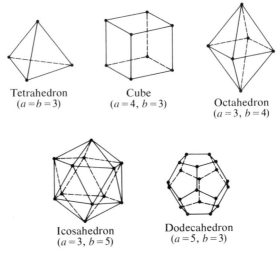

Tetrahedron
$(a = b = 3)$

Cube
$(a = 4, b = 3)$

Octahedron
$(a = 3, b = 4)$

Icosahedron
$(a = 3, b = 5)$

Dodecahedron
$(a = 5, b = 3)$

Figure 1.9   The platonic solids.

Exercises

8. Given a polyhedron (not necessarily regular), let F_n be the number of n-gon faces, and let V_n be the number of vertexes at which exactly n edges meet. Verify the following equations:

(a) $F_3 + F_4 + F_5 + \cdots = F$

(b) $V_3 + V_4 + V_5 + \cdots = V$

(c) $3F_3 + 4F_4 + 5F_5 + \cdots = 2E$

(d) $3V_3 + 4V_4 + 5V_5 + \cdots = 2E$

(e) $(2V_3 + 2V_4 + 2V_5 + \cdots) - (F_3 + 2F_4 + 3F_5 + \cdots) = 4$

(f) $(2F_3 + 2F_4 + 2F_5 + \cdots) - (V_3 + 2V_4 + 3V_5 + \cdots) = 4$

9. Using (e) and (f) of the preceding exercise, show that

$$(F_3 - F_5 - 2F_6 - 3F_7 - \cdots) + (V_3 - V_5 - 2V_6 - 3V_7 - \cdots) = 8$$

Conclude that every polyhedron must have either triangular faces or trivalent vertexes. Similarly, by eliminating F_6 between (e) and (f), deduce that

$$(3F_3 + 2F_4 + F_5 - F_7 - 2F_8 - \cdots) - (2V_4 + 4V_5 + 6V_6 + 8V_7 + \cdots) = 12$$

Therefore every polyhedron must have triangles, 4-gons, or 5-gons. Show that if a polyhedron has no triangles or 4-gons, then it must have at least twelve 5-gons. Is there a polyhedron with this minimal number of 5-gons?

10. F and V play symmetrical roles in Euler's formula. Let two polyhedra be called **dual** if the number of faces of one is the number of vertexes of the other and vice versa. Find the duals of the platonic solids.

11. Find the solutions of (3) when a or b are allowed to equal two. The corresponding polyhedra are called **degenerate**. Draw them.

12. Show that the Euler characteristic of the annulus is zero. Investigate the characteristic of the double annulus, torus, and so forth.

13. Little is known of the life of Diophantus, but the following problem from a Greek collection supplies some information: his boyhood lasted one sixth of his life, his beard grew after one twelfth more, he married after one seventh more, and his son was born five years later. The son lived one half as long as the father, and the father died four years after the son. How long did Diophantus live?

---

§2 CONTINUOUS TRANSFORMATIONS IN THE PLANE

The topology in this book rests on just two fundamental concepts, both introduced informally in §1: continuity (or continuous transformation) and complex. These two notions, one analytical (pertaining to analysis or calculus) and one combinatorial, represent a genuine division of topology into two subjects: point set topology and combinatorial topology, each of which, although undoubtedly possessing close ties with the other, has enjoyed its own development. Point set topology is devoted to a close study of continuity. This study developed from the movement in the nineteenth century to place the calculus finally on a rigorous foundation. Previous generations of mathematicians going back to Newton and Leibnitz themselves had been too busy developing consequences of differentiation and integration, especially in connection with scientific applications, to pay much attention to these foundations. The result of this neglect was that there was not even agreement on the notion of function, much less on limits or continuity. In the nineteenth century several mathematicians resolved to do something about this situation. The leaders of this movement were Abel (1802–1829), Bolzano (1781–1848), Cauchy (1789–1857), and Weierstrass (1815–1897). The result of their activity was the precise formulation of the notions of function, limit, and derivative that are used today. The influence of set theory, the invention of Cantor (1845–1918) in the second half of the century, led to an abstract theory of continuity: point set topology. Today point set topology is an extensive independent field with applications to many parts of mathematics, particularly in analysis.

Combinatorial topology, on the other hand, developed at first as a branch of geometry. The work of Euler (§1) and a number of nineteenth-century geometers on polyhedra is part of this development. However, the foundations of the subject were laid by Poincaré (1854–1912) in a series of papers published around the turn of the century. Poincaré was motivated by problems in analysis, particularly in the qualitative theory of differential equations. We will follow a quasi-historical approach by presenting some of this theory in Chapter Two. Other names associated with the early development of combinatorial topology, whose ideas will be presented here, are Brouwer (1881–1967), Veblen (1880–1960), Alexander (1888–1971), and Lefschetz (1884–1972). The subsequent growth of combinatorial topology has been as extensive as point set topology with applications particularly in geometry and analysis. In addition, combinatorial topology in recent years has developed a strong algebraic flavor, which has led to the founding of new branches of algebra.

In the remainder of this chapter we present the concepts from point set topology, particularly the definition of continuous transformation, that are required in the rest of the book. At first we restrict ourselves to the plane. Ideas once established there will be easily generalized later. In the plane it makes sense to use Cartesian coordinates, so that every point P is associated with a pair of numbers $P = (x, y)$. Given two points P and $Q = (z, w)$, their sum is defined

$$P + Q = (x + z, y + w)$$

and the product of a point P and a real number t is defined by

$$tP = (tx, ty)$$

These are the usual vector operations of addition and scalar multiplication. In addition, the **norm** of the point P is defined by

$$\|P\| = (x^2 + y^2)^{1/2}$$

This is the Euclidean distance from P to the origin. Using the norm we can express the distance between two points P and Q by $\|P - Q\|$.

Continuous transformations were characterized in §1 as transformations that do not involve any ripping or tearing. To put this in a positive way, continuous transformations must preserve the "nearness" of points; that is,

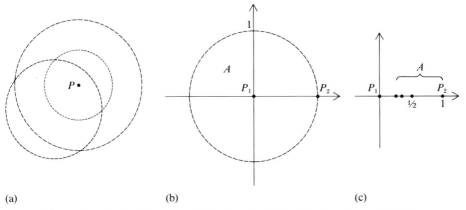

Figure 2.1   (a) A point P with some neighborhoods. (b) P_1 is in A and near A; P_2 is near A but not in A. (c) P_1 is near A; P_2 is near A and in A.

if a point is near a certain set, then the transformed point must be near the transformed set. To make this precise, all that must be done is to define exactly the notion of nearness. In following this approach we are using the ideas of the great analyst F. Riesz (1880–1956).

Definition

Let P be a point in the plane. A **neighborhood** of P is any circular disk (without the boundary circle) that contains P (see Figure 2.1a). Let A be a subset of the plane. The point P is called **near** the set A if every neighborhood of P contains a point of A. If P is near A, we write $P \leftarrow A$.

For example, let A be the open unit disk, that is, the set of points P such that $\|P\| < 1$. Then the points near A include all the points of A plus the points on the boundary circle $\|P\| = 1$ (see Figure 2.1b). On the other hand, if A is the set of points on the x-axis with x coordinates $1, \frac{1}{2}, \frac{1}{3}, \frac{1}{4}, \ldots$, then the only point near A, other than the points already in A, is the origin (see Figure 2.1c).

Exercises

1. Find the points near each of the following sets:

(a) $A =$ the circle $\|P\| = 1$

(b) B = the interval on the x-axis $0 < x < 1$

(c) C = the whole x-axis

(d) D = the points on the x-axis with rational coordinates

(e) E = the points in the plane with rational coordinates

(f) F = the graph of the equation $xy = 1$

(g) G = the points in the plane with integer coordinates

(h) H = the graph of the polar equation $r\theta = 1$

(i) J = the empty set

2. The set A is called **thick** if every point of A is near the set A, even if the point is removed from A. The set A is called **thin** if no point satisfies this condition. Which of the sets of Exercise 1 are thick and which are thin?

3. Prove that if a set A has a point near it but not in it, then A is an infinite set.

4. Let A and B be sets and P a point. Prove that if $P \leftarrow A$ or $P \leftarrow B$, then $P \leftarrow A \cup B$. Conversely, prove that if $P \leftarrow A \cup B$, then either $P \leftarrow A$ or $P \leftarrow B$ or both.

5. Let A and B be sets such that every point in A is near B. Show that every point near A is near B.

6. Let P be a point that is not near the set A. Show that P has a neighborhood disjoint from A.

Continuous transformations can now be defined as the transformations that preserve the relationship of nearness.

Definition

*A **continuous transformation** from one subset D of the plane to another R is a function f with domain D and range R such that for any point $P \in D$ and set $A \subseteq D$, if P is near A, then $f(P)$ is near the set $f(A) = \{f(Q)|Q \in A\}$. In symbols, if $P \leftarrow A$, then $f(P) \leftarrow f(A)$.*

For example, any transformation that preserves the size and shape of circles is continuous, because the definition of nearness is based on circular disks as neighborhoods. Thus in particular, rotations, reflections, and translations of the plane, the usual transformations of Euclidean geometry, are all obviously continuous. It is not too difficult to show that this definition of continuous transformation is equivalent to the familiar $\varepsilon - \delta$ definition from calculus (see Exercises 8 and 9). Therefore functions proved continuous in calculus will be accepted as continuous here without further proof. We will also accept without proof some elementary properties of continuous functions encountered in calculus, for example, the fact that the vector sum of continuous transformations is continuous. The purpose of the following examples is to convince you that continuous transformations are capable of all the contortions intuitively ascribed to them in §1.

Examples

1. *Stretching.* Consider the transformation $s(x, y) = (2x, y)$. This represents a stretching of the plane along the x-axis. All lengths measured parallel to this axis are doubled, while lengths measured parallel to the y-axis are unchanged (see Figure 2.2).

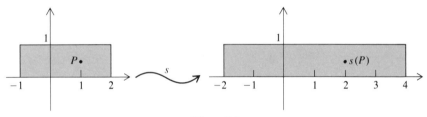

Figure 2.2

2. *Folding.* The transformation $f(x, y) = (|x|, y)$ represents a folding of the plane along the y-axis (see Figure 2.3).

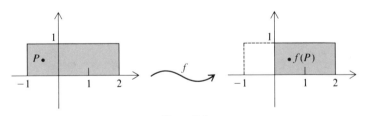

Figure 2.3

3. *Making a quarter annulus.* The transformation $t(x, y) = (x\cos(y),$ $x\sin(y))$ applied to the rectangle $\{(x, y)|1 \leq x \leq 2, 0 \leq y \leq \pi/2\}$ produces a region, four copies of which were used in Figure 1.3 to make an annulus (see Figure 2.4).

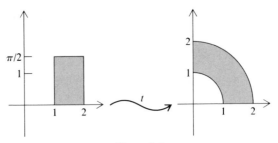

Figure 2.4

4. *Composition.* The composition of two transformations consists of the application of one right after the other. Thus if $s:D \to E$ and $t:E \to F$ are transformations, their composition $t \circ s:D \to F$ is defined by $t \circ s(P) = t(s(P))$. Here is a typical application. Let us find the transformation f^* representing a fold of the plane along the $45°$ line (see Figure 2.5). Instead of working this out directly, we can take advantage of the fact that we already

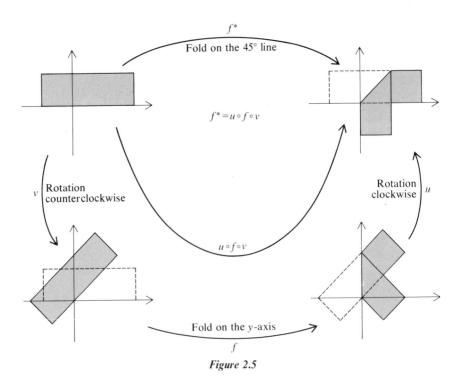

Figure 2.5

know a fold f along the y-axis. First we find the transformation v that rotates the plane $45°$ counterclockwise. This places the $45°$ line at the y-axis. Next apply the transformation f, and finally apply the inverse transformation u, a clockwise rotation of $45°$ that replaces the $45°$ line where it belongs. Then the three-part composition $u \circ f \circ v$ folds the plane on the $45°$ line; in other words, $f^* = u \circ f \circ v$. The rotations v and u are given by the formulas $v(x, y) = (x - y, x + y)/\sqrt{2}$ and $u(x, y) = (x + y, x - y)/\sqrt{2}$ (never mind where these formulas came from; it's the idea that's important). Therefore the formula for f^* is

$$f^*(x, y) = u \circ f \circ v(x, y) = u \circ f\left(\frac{x - y}{\sqrt{2}}, \frac{x + y}{\sqrt{2}}\right) = u\left(\frac{|x - y|}{\sqrt{2}}, \frac{x + y}{\sqrt{2}}\right)$$

$$= \tfrac{1}{2}(x + y + |x - y|, x + y - |x - y|)$$

The following definition explains the use of the word "inverse" in this example.

Definition

*The **identity** transformation I is the transformation leaving each point fixed: $I(P) = P$. A transformation $u: D \to R$ is called **invertible** if there is a transformation $v: R \to D$ so that the composition $v \circ u = I$. The transformation v is called an **inverse** for u. It takes each point as transformed by u and replaces it where it was before u transformed it. In the preceding example, u and v are each inverses of the other: one rotates clockwise, the other counterclockwise. Each undoes the action of the other.*

5. *An infinite stretch.* The transformation $w(P) = P/(1 - \|P\|)$ is continuous as long as its domain is restricted to the unit disk: $D = \{P \mid \|P\| < 1\}$ so that the denominator $(1 - \|P\|)$ is never zero. Then w is a stretching of the disk D onto the whole plane (Figure 2.6).

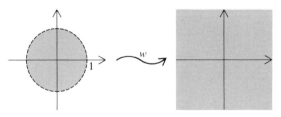

Figure 2.6

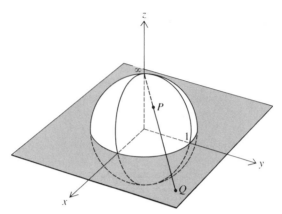

Figure 2.7   Stereographic projection.

6. *Stereographic projection.* This is another infinite stretch. The domain is the sphere $x^2 + y^2 + z^2 = 1$ in space with the north pole removed (see Figure 2.7, where the north pole, the point $(0, 0, 1)$, is marked by the symbol ∞). Consider the transformation s taking each point $P = (x, y, z)$ to the point Q in the (x, y) plane so that the points P, Q, and the north pole lie in a straight line. A little work with similar triangles leads to the result that

$$s(x, y, z) = \left(\frac{x}{1 - z}, \frac{y}{1 - z}, 0 \right)$$

Clearly this definition makes sense for every point on the sphere *except* the north pole. Under this transformation, called **stereographic projection**, the lower hemisphere corresponds to the inside of the unit disk, while the upper hemisphere corresponds to the outside of the unit disk. This is the map projection traditionally used by cartographers to map the polar regions. The north pole itself can be thought of as corresponding to a fictitious "point at infinity" in the plane. This explains the use of the symbol ∞. You can appreciate now why topologists claim that a sphere is nothing but a plane with an extra point added at infinity. These last two examples are a warning: continuous transformations can do things that are impossible for real rubber!

We can now formally define topology.

Definition

A **topological transformation** *is a continuous transformation that has a continuous inverse transformation. In other words, a topological transformation* *is a continuous transformation that can be continuously reversed or undone. Two figures (i.e., sets) in the plane are called* **topologically equivalent** *if there is a topological transformation between them.* **Topology** *is the study of those properties of figures that endure when the figures are subjected to topological transformations. In particular, a* **topological property** *of a figure is a property possessed alike by the figure and all its topological equivalents.*

Topology is therefore defined in terms of the *invertible* continuous transformations. To use arbitrary continuous functions would actually be too general, although this was not mentioned in §1. For example, every set has continuous transformations to one-point sets: the constant functions that are always continuous. It would be ridiculous to study a subject in which all sets were equivalent to a single point. Topological transformations, the reversible continuous transformations, embody much better the intuitive ideas of §1.

We now define a **cell** as any figure topologically equivalent to the (closed) disk $D = \{P \mid \|P\| \leq 1\}$. Two other topological figures of importance are the **path**, any figure topologically equivalent to the line segment $L = \{(x, y) \mid y = 0, 0 \leq x \leq 1\}$; and the **closed path** (or **Jordan curve**), any figure equivalent to the circle $\mathcal{J} = \{P \mid \|P\| = 1\}$.

Exercises

7. Show that the composition of two continuous transformations is continuous.

8. Let $f : D \to R$ be a function. The usual definition of continuity is as follows: for every point P of D and every $\varepsilon > 0$ there is a $\delta > 0$ such that when $\|Q - P\| < \delta$, then $\|f(Q) - f(P)\| < \varepsilon$. Assume that f is continuous in our sense, and prove that f is continuous in the sense of the usual definition.

9. Suppose that the transformation $f : D \to R$ is continuous in the usual sense. Prove that f is continuous in our sense.

10. Let f be any invertible transformation. Show that f is one-to-one, meaning that if $P \neq Q$, then $f(P) \neq f(Q)$.

11. Which of the above examples of continuous transformations are topological? If a transformation is not topological, try to find a large part of its domain such that the function as restricted to that part is topological.

12. Find a formula for continuous transformations to fit the following specifications. Whenever possible, use compositions of the previous examples.

(a) a translation

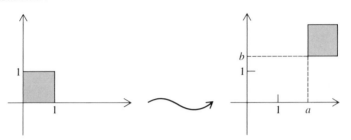

(b) square to a quarter annulus

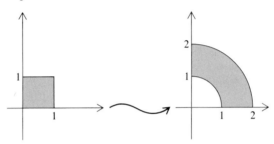

(c) square to a half annulus

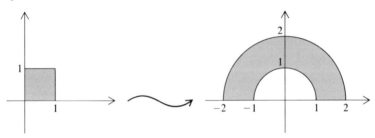

(d) square to a diamond

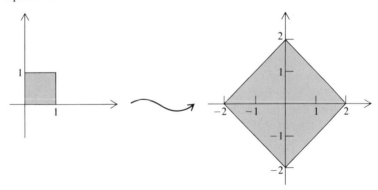

(e) diamond to a square

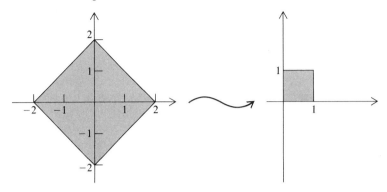

(f) a fold around the line $y = 1$

(g) Stretch a strip to the whole plane.

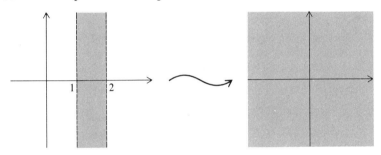

(h) Fold a diamond into a rectangle.

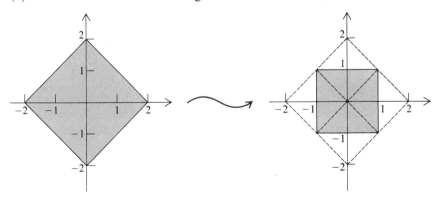

(i) Describe your favorite paper airplane or other origami creation by means of a continuous transformation.

§3 COMPACTNESS AND CONNECTEDNESS

This section presents the fundamental topological properties on which the rest of the book is based. The first of these, compactness, plays a particularly crucial role in the theory. It happens that the proofs of many important theorems in combinatorial topology fall naturally into two parts: a combinatorial part, in which some counting argument is featured, and a point set part, in which limits and continuity are used. For the point set parts we require some means of producing points near given sets when necessary. The concept of compactness suits this purpose exactly because it is designed to ensure that a set has lots of near points. In order to define compactness we require the following definition.

Definition

*Let $\mathscr{P} = \{P_1, P_2, P_3, \ldots\}$ be a sequence of points. The point P is **near** the sequence $\mathscr{P} = \{P_n\}$ if either $P = P_n$ for an infinite number of terms of the sequence or $P = P_n$ only a finite number of times and P is near the set of other values of $\mathscr{P}$.*

Nearness to sequences can be used to characterize continuous functions in the same way as nearness to sets.

Theorem

A function $f : D \to R$ is continuous if and only if whenever the point P is near the sequence $\{P_n\}$ in D, then the point $f(P)$ is near the sequence $\{f(P_n)\}$ in R.

This property of continuity is often easier to use in proofs than the original definition. Its proof is left as an exercise. Now here is the definition of compactness.

Definition

*A set S in the plane is **sequentially compact** if every sequence in S has a near point in S.*

This type of compactness is called sequential in part to distinguish it from other types. We will simply call it compactness until the other types appear (in §41). Compactness is a topological property. Suppose that $f : D \to R$ is a topological transformation from D onto R. Actually in this case it suffices to assume that f is continuous; a stronger assumption is not necessary. In any event, suppose that D is compact. To prove that R is compact, let $\mathcal{Q} = \{Q_n\}$ be a sequence in R. For each Q_n, let P_n be an element of D such that $f(P_n) = Q_n$. Because D is compact, there is a point P in D near the sequence $\mathcal{P} = \{P_n\}$. By continuity the point $f(P)$ is near the sequence $f(\mathcal{P}) = \{\mathcal{Q}\}$ in R, proving that R is compact.

Definition

A set is **bounded** *if it is contained in a rectangle. A set is* **closed** *if it contains all its near points.*

Compact sets are both closed and bounded. These two properties explain somewhat the use of the fanciful term "compact." To prove that compact sets are bounded, consider a set S which is *not* bounded. Then S is not contained in any rectangle. In particular, for each square R_n centered at the origin (see Figure 3.1) there is a point P_n in S but outside R_n. The sequence $\mathcal{P} = \{P_n\}$ has *no* near point (Exercise 1); therefore S is *not* compact. This proves that every compact set is bounded. A similar argument (Exercise 2) proves that compact sets are closed.

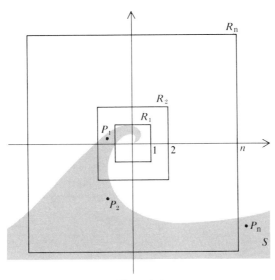

Figure 3.1

Most of the figures that we shall study are compact. To begin, there is the following theorem.

Bolzano-Weierstrass Theorem

Cells are compact.

The proof of this theorem rests on fundamental properties of the real numbers. To give a complete proof would involve too much of a digression. We will prove the following interesting corollary instead.

Theorem

A subset S of the plane is compact if and only if S is closed and bounded.

We have already established that compact sets are closed and bounded. Now we must prove the converse. Therefore suppose that the set S is closed and bounded. Let $\{P_n\}$ be a sequence in S. To prove that S is compact we must find a near point for $\{P_n\}$ in S. Because S is bounded, there is a rectangle R enclosing S. The rectangle R is a cell, hence compact by the Bolzano-Weierstrass theorem. Therefore the sequence $\{P_n\}$ has a near point P in R. Since P is near the sequence $\{P_n\}$, P is also near or in the set S. Then because S is closed, it follows that P must be *in* S. This completes the proof.

Exercises

1. Complete the proof that compact sets are bounded by showing that the sequence $\mathscr{P}$ constructed there has no near points.

2. Let the point P be near the set S. Show that S contains a sequence whose only near point is P. Use this result to prove the first theorem of this section and to prove that compact sets are closed.

3. Which of the following sets are bounded, which are closed, and which are both (i.e., compact)?

   (a) the unit disk including the unit circle

   (b) the unit circle

   (c) the unit disk minus the unit circle

   (d) the x-axis

   (e) the whole plane minus the x-axis

   (f) all points with integer coordinates

(g) the set $\{(1, 0), (\frac{1}{2}, 0), (\frac{1}{3}, 0), (\frac{1}{4}, 0), \ldots\}$

(h) the whole plane

(i) the empty set

4. The property of being closed and bounded is a topological property, namely compactness. Show, however, by examples that neither closedness nor boundedness alone is a topological property.

5. Prove that Jordan curves are compact.

6. For any set S, the **closure** of S consists of S plus all the points near S. Find the closure of each of the sets of Exercise 3. Prove that the closure of any set is closed.

In §1 the disk and the annulus were described as being "of one piece." The formal topological property corresponding to this intuitive phrase is called connectedness.

Definition

A subset S of the plane is **connected** *if whenever S is divided into two non-empty, disjoint subsets A and B ($S = A \cup B$, $A \neq \emptyset$, $B \neq \emptyset$, $A \cap B = \emptyset$), one of these sets always contains a point near the other.*

In other words, a connected set is one that can't be divided into two pieces unless the two pieces are near each other in the sense that one of them contains a point near the other. That this is a topological property is almost obvious, since it is based only on nearness, which is our fundamental topological property. Here is a formal proof. Let $f: D \to R$ be a topological transformation from a connected set D onto a set R. Actually, as with compactness, we need only suppose that f is continuous. To investigate the connectedness of R, let R be divided into two nonempty sets A and B such that $A \cup B = R$ and $A \cap B = \emptyset$. Consider the sets $E = \{P \in D \mid f(P) \in A\}$ and $F = \{P \in D \mid f(P) \in B\}$. These are not empty because f is onto R, and obviously $E \cup F = D$ and $E \cap F = \emptyset$. Since D is connected, one of these sets contains a point near the other; say E contains a point P near F. Then $f(P)$ is in A and by continuity $f(P) \leftarrow B$. This proves that R is connected.

As with compactness, most of the figures we encounter will be connected. To begin, we prove the following theorem.

Theorem

Paths are connected.

It suffices to prove that just one particular path is connected, since both connectedness and the property of being a path are topological properties. Therefore consider the unit interval on the x-axis. Suppose that this interval is divided into two nonempty pieces A and B: $A \cup B = [0, 1]$ and $A \cap B = \varnothing$. To prove connectedness, we must find a point in one of these sets near the other set. The proof uses a technique called the method of bisection. Assume that the endpoints of the interval come from both of the sets A and B. Otherwise if both endpoints were from the same set, we would just confine our attention to some subinterval whose endpoints came from both sets. Bisect the interval into the halves $[0, \frac{1}{2}]$ and $[\frac{1}{2}, 1]$. Then one of these halves has the property that its endpoints come from both sets A and B. Choose this half and bisect it in turn, obtaining two quarters of the original interval. Now one of these quarters contains endpoints from both sets A and B. Continuing in this way we obtain an infinite sequence of subintervals each half the length of its predecessor and each with an endpoint from each set A and B (see Figure 3.2).

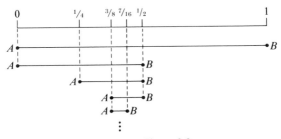

Figure 3.2

Let $\mathscr{P}$ be the sequence of endpoints that belong to A. Then by compactness of the unit interval there is a point P near $\mathscr{P}$. This is the point we are seeking, for since the lengths of the subintervals tend to zero, the point P is also near the sequence $\mathscr{Q}$ of endpoints from the set B. Thus whether P is in A or B it is near the other set, and in this way the proof is concluded.

Exercises

7. Prove that if X and Y are connected and $X \cap Y \neq \varnothing$, then $X \cup Y$ is also connected.

8. Prove that the only connected subsets of the x-axis are intervals.

9. Let f be a continuous real-valued function defined on the closed interval $[a, b]$. Use connectedness and compactness to prove that the range of f, $R = \{y \mid y = f(x) \text{ for some } x \in [a, b]\}$, is a closed and bounded interval also. Deduce from this

(a) the maximum-value theorem: the range of f contains a value M larger than all the other numbers in R

(b) the intermediate-value theorem: if the range of f contains the values c and d, then it must contain all values y between c and d

10. Let S be a plane set such that for any two points of S there is a connected subset of S containing those two points. Prove that S is connected.

11. Prove that cells are connected.

The remaining exercises introduce the few other notions from point set topology required for this book.

---

12. A subset S of the plane is called **open** if every point of S is not near the complement of S. Prove that a point P is in an open set S if and only if P has a neighborhood entirely contained in S.

13. Which of the sets in Exercise 3 are open?

14. For any set S, the **interior** of S, $I(S)$, consists of those points of S that are not near the complement S'. Prove that $I(S)$ is always an open set. Find $I(S)$ for the sets in Exercise 3.

15. Prove that the set S is open if and only if S' is closed.

16. For any set S the **boundary** of S, $b(S)$, consists of those points that are near both S and S'. Find the boundary of each set in Exercise 3. Show that $b(S)$ is the intersection of the closures of S and S'.

17. Show that a set S is open if and only if S does not contain a single point of $b(S)$. Show that S is closed if and only if S contains all of $b(S)$. Use these results to give a second proof for Exercise 15.

18. For any set S prove that every point of the plane is in exactly one of the sets $I(S)$, $I(S')$ and $b(S)$.

---

§4 ABSTRACT POINT SET TOPOLOGY

Nearness is the fundamental concept upon which we have chosen to base all other topological concepts. Thus continuity, compactness, connectedness, closed and open sets, all are defined using nearness. At the same time, nearness is itself based on the notion of neighborhood. This suggests that we can study topology on any set by first choosing a class of subsets to serve as the neighborhoods in that set and then carrying over unaltered all the definitions of the other notions from the topology of the plane. This is what abstract point set topology is all about.

Definition

*A **topological space** is a set $\mathscr{S}$ together with the choice of a class of subsets* **N** *of $\mathscr{S}$ (each of which is called **neighborhood** of its points) such that*

(a) *Every point is in some neighborhood.*

(b) *The intersection of any two neighborhoods of a point contains a neighborhood of that point.*

Examples

1. *Usual topologies.* For the plane the neighborhoods are usually chosen to be the open disks. It is easy to verify that this choice satisfies all the above requirements. The resulting topological space is the space we have been studying all this chapter and will continue to study through Chapter Three. Now for every abstract mathematical concept there are certain *canonical examples:* the examples upon which the abstract theory is modeled, whose properties become theorems in the abstract theory, whose generalization led to the abstract theory in the first place. In point set topology the crucial canonical example is the plane. In trying to master the general theory you should always keep this particular example in mind, because its properties have largely directed the development of the whole subject.

Another canonical example is the usual topology on the real line. Here the neighborhoods are taken to be the open intervals. As with the plane, one easily verifies that this choice satisfies the requirements for a topology. This is the topological space whose study occupies much of the calculus.

Finally we should mention the usual topology in space whose neighborhoods are the open spherical balls about each point.

2. *The discrete topology.* Any set $\mathscr{S}$ becomes a topological space if we allow *all* nonempty sets to be neighborhoods. This extreme topology is often a useful example with which to test conjectures.

3. *The indiscrete topology.* Any set $\mathscr{S}$ becomes a topological space if we stipulate that the *only* neighborhood is the whole set $\mathscr{S}$ itself. Here, in contrast with the discrete topology where as many neighborhoods as possible are permitted, the minimum number of neighborhoods is allowed consistent with the requirements of the definition of topological space. Both the discrete and indiscrete topology have unusual properties.

4. *The finite topology.* Still another topology that can be defined on any set $\mathscr{S}$ arises by choosing as neighborhoods all subsets of $\mathscr{S}$ with finite complement. You should verify that this example, as well as the others (of course), meets the requirements of the definition of topological space.

5. *The sector topology.* Here is an *unusual* topology on the plane. The neighborhoods are sectors of angles symmetric about the vertical line through their vertexes and including the portion of this line extending below the point (see Figure 4.1).

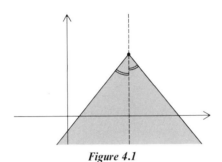

Figure 4.1

6. *The either/or topology.* This is another unusual topology in the plane. The neighborhoods are *either* disks that do not intersect the x-axis *or* the whole plane minus a finite number of points of the x-axis.

7. *Compactification of the plane.* Let $\mathscr{S}$ be the plane plus an extra point called ∞. Neighborhoods are the usual disks in the plane plus neighborhoods of ∞ consisting of ∞ together with the outside of the closed disks in the plane. This space is topologically equivalent to a sphere.

Definition

*Let $\mathscr{S}$ be a topological space. Let A be a subset of $\mathscr{S}$ and P be a point of $\mathscr{S}$. P is **near** A, written $P \leftarrow A$, if every neighborhood of P contains a point of A. P is **near** a sequence $\{P_n\}$ if either $P = P_n$ infinitely often or P is near the set of other values of the sequence. (Instead of nearness, most topology texts use the language of limit points. See Exercise 14.)*

Exercises

1. Verify that all the above examples fill the defining conditions of a topological space.

2. Let $\mathscr{S}$ be any topological space. Prove that nearness has the following properties:

(a) If a subset A has a near point, then A is nonempty.

(b) If A and B are subsets of $\mathscr{S}$ such that $A \subseteq B$, then $P \leftarrow A$ implies $P \leftarrow B$.

(c) If $P \leftarrow A \cup B$, then either $P \leftarrow A$ or $P \leftarrow B$.

3. Show that if the set $\mathscr{S}$ has the discrete topology, then no subset A has any outside near points. State and prove the corresponding result for the indiscrete topology.

4. Find the near points in the finite topology for each of the sets listed in Exercise 3 of §3. Repeat this exercise for the sector topology and the either/or topology.

Definition

Let $\mathscr{S}$ be any topological space. A subset A of $\mathscr{S}$ is called **closed** *if A contains all its near points,* **open** *if every point in A is not near the complement of A.* **(sequentially) compact** *if every sequence in A has a near point in A, and* **connected** *if whenever A is divided into two nonempty disjoint parts then one of these parts contains a point near the other. Associated with each set A are the* **interior** *of A consisting of all points of A not near A', the* **closure** *of A consisting of A plus all points near A, and the* **boundary** *of A consisting of the points near both A and A'.*

Depending on the topological space, these concepts can have very unusual, apparently even paradoxical, interpretations. For example, in the discrete topology no set A ever has a near point in A'; therefore every set is both open and closed (check this!). On the other hand, no set is connected (except one point sets), and no set is compact (except finite sets). Abstract point set topology takes some getting used to! Of course these results are *not* paradoxical. They follow logically from the definitions and axioms. Everything depends on the choice of neighborhoods. In the discrete case, by choosing so many neighborhoods we make it dreadfully difficult for a set to get near an outside point, in fact impossible. We should picture the points of a discrete space as quite separated from each other; then the properties of the space seem less strange.

Certain relationships between these fundamental notions hold in any topological space, while others hold only in certain spaces. Examples of both types are given in the following exercises.

Exercises

5. Find examples of plane sets that are closed for the indiscrete, finite, sector, and either/or topologies. In each case try to find a description of the typical closed set. Do the same with open, compact, and connected sets.

6. Prove the following theorems for any topological space:

(a) A point P is in an open set A if and only if P has a neighborhood entirely contained in A.

(b) The union of any number of open sets is open.

(c) The intersection of a finite number of open sets is open.

Show by an example that the intersection of an infinite number of open sets need not be open.

7. Prove in any topological space that a set is open if and only if its complement is closed.

8. A set that is both open and closed is called **clopen**. A topological space always contains at least two clopen sets: the whole space and the empty set. Prove that a topological space is connected if and only if it contains no other clopen sets. Prove that a space is connected if and only if it cannot be divided into nonempty, disjoint open parts.

9. For each of the plane topologies usual, discrete, indiscrete, finite, sector, and either/or, consider each of the following statements:

(a) A set consisting of a single point is closed.

(b) A set consisting of a circle is closed.

(c) Compact sets are closed.

(d) Compact sets are bounded.

In each case either prove the statement true in the topology or find a counterexample.

This introduction to point set topology is completed by the following definition.

Definition

Let $\mathcal{S}$ and $\mathcal{T}$ be topological spaces. A transformation $f: \mathcal{S} \to \mathcal{T}$ is **continuous** *if for any point P of $\mathcal{S}$ and subset A of $\mathcal{S}$, $P \leftarrow A$ implies $f(P) \leftarrow f(A)$. A* **topological transformation** *is a continuous invertible transformation whose inverse is also continuous. Two sets are* **topologically equivalent** *if there is a topological transformation between them.* **Topology** *is the study of properties of sets that are preserved when the sets are subjected to topological transformations.*

Exercises

10. Prove that the composition of two continuous transformations is continuous.

11. Let $\mathcal{S}$ be a discrete space. Show that *every* function from $\mathcal{S}$ to itself is continuous. What is the corresponding result for the indiscrete topology?

12. Prove that compactness and connectedness are topological properties.

13. Prove that Example 7 is topologically equivalent to a sphere with the usual topology.

14. Let A be a subset of a topological space $\mathcal{S}$. A point P of $\mathcal{S}$ is called a **limit point** of A if every neighborhood of P contains a point of A other than P itself. Show that if P is a limit point of A, then P is near A, but conversely a point near A need not be a limit point of A.

Notes. The informal discussion of combinatorial topology begun in this chapter will be gradually supplanted by a more and more formal treatment over the course of this book. Naturally it is important to begin with an informal understanding. There are many excellent treatments of topology from an informal point of view. The chapters on topology in the books by Courant and Robbins [6] and Kasner and Newman [15] are good examples. Especially fine is the **picture essay** in the Life Science Library [3]. These references are concerned with the popular aspects of topology. More formal treatments comparable with this book can be found in the books of Blackett [4] and Frechet and Fan [8]. For point set topology a standard reference is Kelley [16]. Our treatment based on nearness was suggested by Cameron, Hocking, and Naimpally [5].

two

Vector Fields

§5 A LINK BETWEEN
ANALYSIS AND TOPOLOGY

Poincaré's first pioneering work in topology grew out of his research into systems of differential equations. Ever since, applications to differential equations have played an important role in topology, especially algebraic and combinatorial topology. It may seem surprising at first that such superficially different subjects as topology and differential equations should be related, but in mathematics such startling connections are common. A link between the two is the concept of a vector field. A **vector field** V on a subset D of the plane is a function assigning to each point P of D a vector in the plane with its tail at P. Intuitively we can think of V as giving the velocity of some substance that is presently in D but in a state of agitation, like water in a bathtub.

Placement of the vector $V(P)$ with its tail at P is mainly for dramatic effect, to aid in visualizing the vector field. As usual with vectors, the only

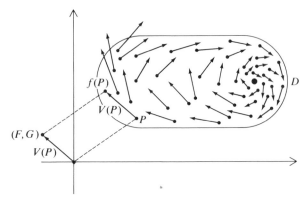

Figure 5.1   A vector field.

essential qualities of the vector $V(P)$ are its length and direction. For practical purposes it is usually more convenient to place all vectors with their tails at the origin. Then $V(P)$ may be described by the coordinates of its head,

$$V(P) = (F(x, y), G(x, y)) \qquad (1)$$

where F and G are real-valued functions of $P = (x, y)$ (see Figure 5.1). The vector field V is called **continuous** when this function (1) is continuous as defined in Chapter One. If we let $f(P)$ stand for the point at the head of the vector $V(P)$ when it is placed with its tail at P, we obtain a transformation on D, the vector sum of P and $V(P)$:

$$f(P) = P + V(P) = (x + F(x, y), y + G(x, y)) \qquad (2)$$

In view of the theorem that the sum of continuous transformations is continuous, f will be continuous if V is a continuous vector field. Conversely, supposing that a transformation f is given on D, then the vector field V can be defined by letting $V(P)$ be the vector from the point P to the point $f(P)$. The vector field V will be continuous if the transformation f is continuous. Clearly the study of vector fields on a set D coincides with the study of continuous transformations of the set.

Vector fields have many important applications. The force fields arising from gravitation and electromagnetism are vector fields; the velocity vectors of a fluid in motion, such as the atmosphere (wind vectors), form a vector field; and gradients, such as the pressure gradient on a weather map or the height gradient on a relief chart, are vector fields. These examples are usually studied from the point of view of differential equations. A vector field (1) determines a system of differential equations in the two unknowns x and y.

These variables are taken to represent the position of a moving point in the plane dependent on a third variable, the time t. The system of differential equations takes the form

$$\begin{cases} x' = F(x, y) \\ y' = G(x, y) \end{cases}$$ (3)

where the differentiation is with respect to t. Such a system is called **autonomous** because the right-hand sides are independent of time. Autonomous systems are of particular interest because the fundamental laws of nature are believed to be independent of time. A solution of the system (3) consists of two functions expressing x and y in terms of t. These may be considered the parametric equations of a path in the plane: the path of a molecule of gas or liquid, the orbit of a planet or an electron, or the trajectory of a marble rolling down a hill, depending on the application. The original vector field $V(P)$ gives the tangent vector to the path of motion at the point $P = (x, y)$.

Examples

1. Consider the vector field $V(x, y) = (2, 1)$. All the vectors are equal, so the corresponding transformation, $f(x, y) = (x + 2, y + 1)$, is a translation (see Figure 5.2a). The resulting system of differential equations, $x' = 2$, $y' = 1$, has solutions $x = 2t + h$, $y = t + k$. The solution paths (dashed in Figure 5.2a) are the family of straight lines of slope $\frac{1}{2}$.

2. Consider the vector field $V(x, y) = (-y, x)$. Some of the vectors are sketched in Figure 5.2b. The corresponding transformation $f(x, y) = (x - y, x + y)$ is a rotation of $45°$ counterclockwise combined with a stretching from

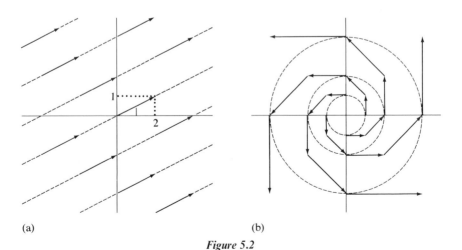

(a)   (b)

Figure 5.2

the origin by a factor of $\sqrt{2}$ (see Example 4 of §2). The system of differential equations, $x' = -y$, $y' = x$, has the solutions $x = k \cos(t)$, $y = k \sin(t)$. The solution paths are the circles centered at the origin. They are tangent at every point to the vector field V.

These three mathematical objects—transformation, vector field, and system of differential equations—are essentially the same. Any one determines the other two, and all are continuous if any one is. Thus vector fields link topology and analysis. The relationship can be exploited in two ways: vector fields and differential equations can be used as tools in topology, and topological ideas enter the field of differential equations. In the next section we see an application of the first sort. The Brouwer fixed point theorem, a theorem of topology, will be proven using vector fields. The remainder of the chapter, on the other hand, is devoted to the topological aspects of differential equations.

Exercises

1. Draw the vector fields corresponding to the examples of continuous transformations given in §2.
2. For each of the following vector fields sketch some of the vectors and describe roughly the corresponding transformation. If possible, solve the corresponding system of differential equations. The solutions will usually be simple exponential, trigonometric, or hyperbolic functions in these examples. Sketch some of the solution paths.

(a) $V(x, y) = (1, y)$ (b) $V(x, y) = (x, y)$

(c) $V(x, y) = (x, -y)$ (d) $V(x, y) = (4x, y)$

(e) $V(x, y) = (-y, 1)$ (f) $V(x, y) = (y, x)$

(g) $V(x, y) = (-y, 4x)$

---

§6 SPERNER'S LEMMA AND THE BROUWER FIXED POINT THEOREM

Let f be a continuous transformation of a set D into itself. Intuitively f is some sort of twisting, stretching, and crumpling in which the transformed set is eventually replaced within the original set (Figure 6.1). It may happen that some point P is brought back to its original position. In effect P is not

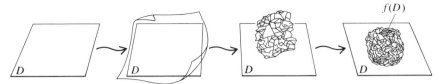

Figure 6.1   Continuous transformation of a cell into itself.

moved, $f(P) = P$, and P is called a **fixed point** of f. If it happens that every continuous transformation, no matter how violent, of D into D has a fixed point, then we say that D has the **fixed point property**. The fixed point property is a topological property. To prove this, let $u:D \to R$ be a topological transformation from a set D with the fixed point property to a set R. Let g be a continuous transformation of R into R. To find a fixed point for g, consider the composition $f = v \circ g \circ u$, where v is the inverse transformation $v:R \to D$. This composition is a continuous transformation of D to D. By hypothesis, f has a fixed point P in D, $v(g(u(P))) = f(P) = P$. This means that $g(u(P)) = u(P)$, so that $u(P)$ is a fixed point for g. This proves that R has the fixed point property.

The following theorem is the most important result on fixed points in the plane.

Brouwer's Fixed Point Theorem

Cells have the fixed point property.

Thus the transformation of Figure 6.1, as well as every other continuous transformation of a cell, has a fixed point! The proof is quite long but involves many important ideas. It divides neatly into a combinatorial part (Sperner's lemma) and a point set part (topological lemma), a pattern of proof already mentioned in §3. We begin with the combinatorial part.

Consider a cell in the form of a Euclidean triangle, and let this triangle be further divided into subtriangles (Figure 6.2). Later we will study much more complicated situations, but for the moment we may as well assume that all the triangles involved are Euclidean triangles with straight sides. Then we further assume this subdivision of the cell to be a *triangulation* in the following technical sense.

Definition

A **triangulation** *of a triangle D (or any other polygon) is a division of D into a finite number of triangles so that each boundary edge of D is the edge of just one*

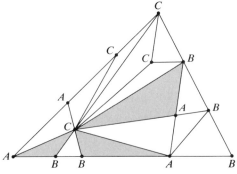

Figure 6.2   A Sperner labeling.

triangle of the subdivision, and each edge in the interior of D is an edge of exactly two triangles of the subdivision.

A triangulation is a type of complex (see §1). The vertexes of the triangulation are now to be labeled with the letters A, B, and C. First the three vertexes of the original triangle are labeled, each with a different label. Then the vertexes on the boundary of the original triangle are labeled (Figure 6.2), subject to the restriction that along the side of the triangle already labeled AB only A's and B's may be used, only B's and C's are to be used along the side already labeled BC, and only A's and C's are to be used along the remaining side (labeled AC). Finally the vertexes inside the triangle can be labeled any old way. A labeling satisfying these conditions is called a **Sperner labeling**. The combinatorial part of the proof of the Brouwer fixed point theorem is contained in the following lemma.

Sperner's Lemma

At least one subtriangle in a Sperner labeling receives all three labels: A, B, and C.

This mysterious result may seem far from fixed points, but combined with an application of compactness it will yield Brouwer's theorem. The triangles that receive all three labels are called **complete triangles**. We shall actually prove that the number of complete triangles is *odd*. In Figure 6.2 the three complete triangles are shaded.

Consider first the analogous problem in one dimension: a single edge labeled AB, subdivided into segments, and the vertexes labeled with A's and

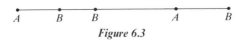

Figure 6.3

B's only. The question here is whether there are any **complete segments**: segments with both labels A and B (see Figure 6.3). We shall prove that the number b of complete segments is odd. To do this let us count the number of vertexes labeled A in the following manner: let a be the number of segments labeled AA. Now the segments of this type have two A vertexes, while the complete segments have only one A vertex. Other types of segment have no A vertexes, so the total number of A vertexes, counted segment by segment, is $2a + b$. But wait! In this total, the A vertexes inside the original segment have been counted twice since they belong to two subsegments. Letting c be the number of internal A vertexes, we have actually counted $2c + 1$ vertexes. Thus $2a + b = 2c + 1$. This proves that b is odd.

Returning to two dimensions, let b be the number of complete triangles in a given Sperner labeling. To prove that b is odd, we count the number of edges labeled AB inside and on the triangle in the following way: let a be the number of triangles whose labels read ABA or BAB. Now the triangles of these two types have two edges labeled AB, while the complete triangles have one edge labeled AB. Other types of triangles have no edges labeled AB, so the total number of these edges, counted triangle by triangle, is $2a + b$. But wait! In this total, the edges inside the original triangle are counted twice since they belong to two triangles. Letting c be the number of edges labeled AB inside the triangle, we have actually counted $2c + d$ edges, where d is the number of edges labeled AB on the outside of the triangle. Thus $2a + b = 2c + d$. According to the one-dimensional result of the preceeding paragraph, d is odd, and therefore b is odd as well. This completes the proof.

Exercises

1. Investigate the examples of continuous transformations given in §2 for fixed points.

2. Find a plane set that does not have the fixed point property.

3. Prove Sperner's lemma in three dimensions: let a tetrahedron, whose vertexes have been labeled A, B, C, and D be divided into subtetrahedra with vertexes labeled so that only three labels appear on each face of the original tetrahedron. Then at least one subtetrahedron has all four labels.

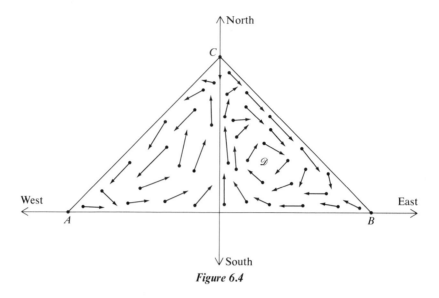

Figure 6.4

For the proof of Brouwer's theorem, consider the triangle $\mathcal{D}$ shown in Figure 6.4. We shall show that this triangle has the fixed point property. Since the triangle is a cell and the fixed point property is a topological property, it will follow immediately that all cells have this property. Basing the proof on a particular figure is a well-established geometric tradition. In Euclidean geometry, for example, one always chooses the figure in some convenient relation to the coordinate axes so that some special property of the figure can be used in the proof. In this case the proof will depend not only on the particular orientation of $\mathcal{D}$ to the coordinate axes but also on the fact that $\mathcal{D}$ is a triangle.

Let f be a continuous transformation of $\mathcal{D}$ into itself, and let V be the corresponding vector field. In vector field terms, we seek a point P where $V(P) = 0$. This will be a fixed point for f. Intuitively here is the argument: since f maps $\mathcal{D}$ into itself, the vectors V all lie entirely inside $\mathcal{D}$. On this account these vectors must point in all kinds of different directions. For example, the vector at the vertex C of Figure 6.4 must point south (more or less), while those at A and B point northeast and northwest, respectively. In the course of the proof we will find many more examples of triangles inside $\mathcal{D}$ with this same property: at the vertexes of these triangles one vector points south, one northeast, and one northwest. By producing an infinite number of such triangles and applying compactness, we will find a point P near the set of points with south-pointing vectors. At the same time P will be near the set of points with northeast-pointing vectors and near the set of points with northwest-pointing vectors. By continuity, the vector $V(P)$ will point all three directions simultaneously. It will follow that $V(P) = 0$, completing the proof. We now have a few details to fill in!

For the purposes of this proof we decide that *there shall be only three directions: south, northeast, and northwest* determined by the lower half plane and first and second quadrants, respectively. A given vector is placed with its tail at the origin and its *direction* determined by the location of its head. Figure 6.5 illustrates these three directions. Vectors pointing south in this sense actually range from east to west in the usual sense. Every nonzero vector points at least one of these directions. There are three situations where vectors point two different directions: vectors pointing due east, north, or west in the usual sense. For example, a vector pointing due north is considered to point both northeast and northwest in our sense. Note, however, that only the zero vector points all three directions simultaneously.

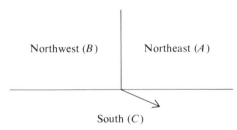

Northwest (B)   Northeast (A)

South (C)

Figure 6.5   A south-pointing vector.

Let $\mathcal{D}$ be triangulated, and label each vertex of the triangulation according to the direction of the vector V at that vertex: A for points with northeast vectors, B for points with northwest vectors, and C for points with south-pointing vectors. To the points with vectors of ambiguous direction (due north, east, and west) we assign the label A for definiteness in the first two cases and the label B in the last case. The result is a Sperner labeling. Therefore by Sperner's lemma, one of the triangles of the triangulation has a complete set of labels, and the vectors of this triangle point all three directions (see Figure 6.6).

It is important to point out that $\mathcal{D}$ contains complete triangles with arbitrarily small sides. Figure 6.7 shows a sequence of triangulations of $\mathcal{D}$ in which the size of the triangles decreases toward zero. Each of these triangulations contains a complete triangle. All these ingredients can now be assembled in a complete proof. The remaining argument is purely topological. Because of its importance it is stated as a separate lemma.

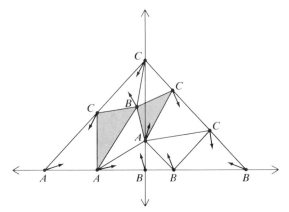

Figure 6.6   A Sperner labeling obtained from a vector field.

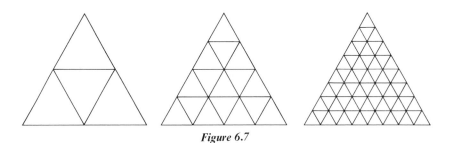

Figure 6.7

Topological Lemma

Let a continuous vector field V be defined on the compact set $\mathcal{D}$. Imagine each point of $\mathcal{D}$ labeled according to the direction of the vector field V. If $\mathcal{D}$ contains complete triangles of arbitrarily small size, then there is a point P in $\mathcal{D}$ where $V(P) = 0$.

By hypothesis, $\mathcal{D}$ contains a sequence of complete triangles with vertexes P_n, Q_n, and R_n labeled *A*, *B*, and *C*, respectively, such that the length of the sides of these triangles tends to zero. Thus there are three sequences of vertexes $\mathcal{P} = \{P_n\}$, $\mathcal{Q} = \{Q_n\}$, and $\mathcal{R} = \{R_n\}$ consisting only of points with northeast-, northwest-, and south-pointing vectors, respectively. *By compactness* there exists a point P of $\mathcal{D}$ near $\mathcal{P}$. Since the sides of the triangles tend to zero, it follows that P is also near $\mathcal{Q}$ and near $\mathcal{R}$. *By continuity* the vector $V(P)$ is near the sequence of vectors $V(\mathcal{P})$ as well as the sequences $V(\mathcal{Q})$ and $V(\mathcal{R})$. The vectors $V(\mathcal{P})$ all point northeast. Since the set of northeast-pointing vectors, namely the first quadrant, is closed, it follows that

$V(P)$ is northeast-pointing too. Similarly, $V(P)$ is northwest-pointing and south-pointing. Since the zero vector is the only vector pointing all three directions, $V(P) = 0$, completing the proof of Brouwer's theorem and the lemma.

It is worth remarking that the topological portion of the proof of Brouwer's theorem depends only on the compactness of the set $\mathscr{D}$ and continuity of the vector field. This is in contrast with the combinatorial portion, where the triangular nature of $\mathscr{D}$ is crucial. To emphasize the generality of the topological argument, we restate it in a form suitable for application later.

Topological Lemma

Let $\mathscr{D}$ be any compact set with a continuous vector field V. If V is never zero on $\mathscr{D}$, then there is a constant $\varepsilon > 0$ such that every complete triangle with vertexes in $\mathscr{D}$ has a side of length greater than ε.

Exercises

4. Explain why the labeling used in the proof of Brouwer's theorem is a Sperner labeling.
5. Outline a proof of Brouwer's fixed point theorem in three dimensions: the solid enclosed by a tetrahedron has the fixed point property.
6. Justify the restatement of the topological lemma given at the end of this section.
7. Prove that a triangle is a cell.

---

§7 PHASE PORTRAITS AND THE INDEX LEMMA

Consider the system of differential equations

$$\frac{dx}{dt} = F(x, y)$$

$$\frac{dy}{dt} = G(x, y)$$

(1)

determined by a continuous vector field $V(x, y) = (F(x, y), G(x, y))$ in some region $\mathscr{D}$ of the plane. Regarding the independent variable t as a parameter, the solutions form a family of *directed paths* in the plane, called **integral paths** of the system, which are tangent at each point P, through which they pass, to the vector $V(P)$ at that point. According to the basic existence and uniqueness theorems for this system (which we shall not prove), exactly one of these integral paths passes through each point P at which $V(P)$ is not zero. The picture formed by these paths is called the **phase portrait** of the system of differential equations. An example is given in Figure 7.1. Poincaré discovered that the nature of the phase portrait is determined by the exceptional points P, called **critical points**, where $V(P) = 0$ and around which the integral paths gather. In Figure 7.1 there are four critical points. At the top is a **center** characterized by the closed integral curves that swirl around it. No integral path passes through the center. At the bottom are two critical points called **nodes** characterized by the fact that all the integral paths near these points end there. The difference between the two nodes can be expressed by saying that one is stable and the other is unstable. In general, a critical point is called **stable** if one can find a cell surrounding the point from which no integral paths exit. The idea behind this definition is that a point put down near a stable critical point and constrained to follow the integral paths will stay near the critical point. Thus the node where the arrows are directed toward the point is the stable one. The center is also a stable critical point. In between the center and the nodes is a **saddle point**, where exactly four integral paths meet, two beginning and two ending. The saddle point is unstable.

The most important characteristics of the phase portrait are the number and arrangement of the critical points, the pattern of the integral paths

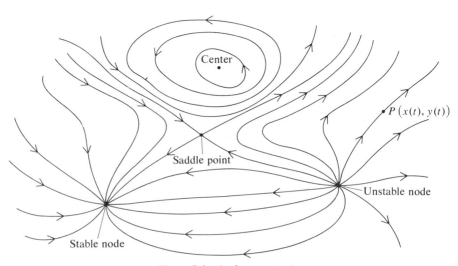

Figure 7.1 A phase portrait.

about each point, and the stability or instability of the critical points. These properties are part of a purely topological theory of differential equations developed by Poincaré. For if we imagine the portion of the plane represented in Figure 7.1 cut out of a sheet of rubber as in §1 and subjected to a continuous transformation, then, although the shape and length of the paths could change, the number and nature of the critical points cannot change; nodes remain nodes (stable or unstable); saddle points remain saddle points; and so forth. These properties, to which the rest of this chapter is devoted, are topological properties of the phase portrait, topological properties of the system of differential equations.

Examples

In a few cases the integral paths of a vector field (or corresponding system of differential equations) may be determined by solving the single differential equation

$$\frac{dy}{dx} = \frac{G(x, y)}{F(x, y)} \tag{2}$$

for y as a function of x, instead of solving the system (1) for x and y in terms of t. For example, consider the vector field $V(x, y) = (x, 2y)$. In this case equation (2), $dy/dx = 2y/x$, is separable. Separating and integrating we obtain the solutions $y = Kx^2$. These integral curves (parabolas) are plotted for a few values of K in Figure 7.2 along with a few of the vectors V themselves from which the correct direction of motion along the integral paths can be determined. The only critical point is an unstable node at the origin.

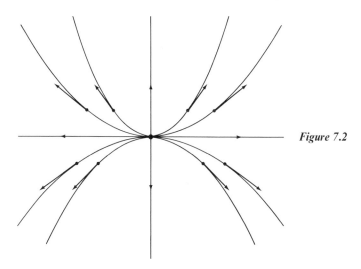

Figure 7.2

Exercises

1. For each of the following vector fields use the differential equation (2) to sketch the phase portrait, and classify the critical points by stability and type.

(a) $V(x, y) = (x, y)$

(b) $V(x, y) = (y, -x)$

(c) $V(x, y) = (2y, -x)$

(d) $V(x, y) = (y, x)$

(e) $V(x, y) = (x, -y)$

(f) $V(x, y) = (x^2 - 1, 2xy)$

(g) $V(x, y) = (1 - x^2, 2xy)$

(h) $V(x, y) = (x^2 - 1, y)$

(i) $V(x, y) = (y, 1 - x^2)$

(j) $V(x, y) = (x^2 - y^2, 2xy)$

(k) $V(x, y) = (x, x^2 - y)$

2. Let the continuous vector field V be defined on the cell $\mathscr{D}$. Supposing that V has the property that all its vectors lie inside $\mathscr{D}$, show that V must have a critical point in $\mathscr{D}$.

Developments in succeeding sections are based on a generalization of Sperner's lemma: the index lemma. Consider a cell in the form of a polygon of any number of sides. Let the cell be triangulated and the vertexes labeled in any manner whatsoever with the labels A, B, and C. Figure 7.3 gives an example. Two quantities are associated with the resulting figure. The **content** C is defined to be the number of complete triangles counted *by orientation*. This means that each triangle counts plus one if its labels read ABC in a

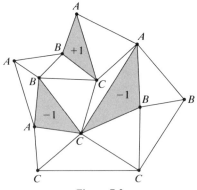

Figure 7.3

counterclockwise direction around the triangle, but counts minus one if the labels read *ABC* clockwise around the triangle. This is in accord with a general convention in mathematics that counterclockwise means plus, clockwise minus. The **index** *I* is defined to be the number of edges labeled *AB* around the boundary of the polygon counted by orientation, meaning that an edge counts plus one if it reads *AB* counterclockwise around the polygon, and minus one if it reads *AB* clockwise around the polygon. In Figure 7.3 both index and content are minus one.

Index Lemma

The index equals the content.

For the proof let *S* be the number of edges labeled *AB* on and inside the polygon counted in the following way: each triangle is considered apart from all the others and its edges *AB* counted plus or minus by orientation. The proof is completed by showing that $C = S = I$. To prove the first equality, consider a complete triangle of type ⟨triangle with vertices C top, A bottom-left, B bottom-right⟩. This has just one edge of positive orientation and so contributes one to *S*. At the same time as a complete triangle it counts one as part of *C*. As a second example consider a triangle of type ⟨triangle with vertices B top, A bottom-left, A bottom-right⟩. This has two edges *AB*, one each of positive and negative orientation, and so contributes a total of zero to *S*. It also contributes zero to *C*, since it is not complete. By surveying all the possible types of triangle, one verifies that each triangle is counted in the same way in *S* as in *C*; therefore $C = S$.

On the other hand, consider an individual edge labeled *AB*. If this edge is inside the polygon, then it is an edge in two triangles in one of which it counts plus one while in the other it counts minus one. These edges then contribute nothing to *S*. But edges on the boundary of the polygon are counted only once in *S*, and then just as they are for the index *I*. Therefore $S = I$.

Exercises

3. Deduce Sperner's lemma from the index lemma.

4. In the definition of the index, a special role is played by the edges *AB*. Show that the same values would result if the edges *BC* or *CA* were used instead.

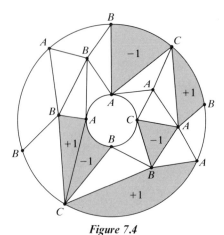

Figure 7.4

5. Consider an annulus triangulated and labeled as in Figure 7.4. The content may be defined as for cells, while there are now two indexes: one for the outside boundary I_1 and one for the inside I_2. Prove that $C = I_1 - I_2$.

§8 WINDING NUMBERS

Consider a continuous vector field V on a closed path γ. Suppose that V is never zero on γ. Starting at a fixed point Q on γ, imagine a variable point P traversing the path γ once in the counterclockwise direction returning to Q. Two examples are given in Figure 8.1. The vector $V(P)$, which is what we are interested in, starting at $V(Q)$ will wiggle about during the trip around γ and return to the position $V(Q)$. During the journey $V(P)$ will make some whole number of revolutions. For example, the vector field V makes a single clockwise revolution on the curve γ_1 of Figure 8.1. Counting these revolutions positively if they are counterclockwise, negatively if they are clockwise, the resulting algebraic sum of the number of revolutions is called the **winding number of V on γ** and denoted $W(\gamma)$. Thus, for example, $W(\gamma_1) = -1$. In contrast on γ_2, after some preliminary hesitation (between Q and T) the vector $V(P)$ makes two counterclockwise revolutions (between T and V) followed by a clockwise revolution before returning to $V(Q)$. Therefore $W(\gamma_2) = +1$. The figure shows only an occasional vector, of course. You must imagine the remaining vectors filled in in a continuous manner. Intuitively the winding number $W(\gamma)$ may be thought of as a measure, although a very crude one, of the activity of the vector field on γ. The winding number

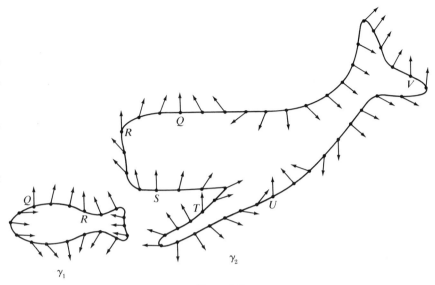

Figure 8.1

will be our main tool in the study of differential equations. As we shall see, it also contains information on the activity of the vector field *inside* γ.

This definition of winding number will have to be made more precise, however, before it becomes useful. Here is an alternative method for computing the winding number, suggested by Poincaré himself. Instead of keeping track of the vector $V(P)$ as P traverses the path γ, we watch just one particular direction of our choosing and record only the times that the vector $V(P)$ points in that direction. If the vector $V(P)$ passes through the direction going counterclockwise, we count plus one; if it passes through the direction going clockwise, we count minus one; and we count zero if the vector only comes up to our chosen direction and then retreats back the way it came. For the distinguished direction we choose due north. In Figure 8.1 the places have been marked where the vector $V(P)$ points due north. On $γ_1$ this occurs twice at Q and at R. At R the vector $V(P)$ is passing through the vertical clockwise, while at Q it approaches the vertical from the right but reverses direction at Q and retreats on the same side. Counting minus one at R, zero at Q, the total winding number is -1. On $γ_2$ there are six occasions when the vector $V(P)$ points north: R, T, and U each count plus one, Q and S count minus one, and V counts zero. The total winding number is $+1$. Note that trouble develops if the vector field oscillates infinitely often about due north. Something will be done about this difficulty shortly.

This method of reckoning the winding number can be refined still further as follows. First partition γ by choosing a finite number of points $\{P_i\}$

dividing γ into a number of edges. The points of subdividion P_i are then labeled according to the direction of the vectors $V(P_i)$ using the same conventions regarding the labels ABC as in the proof of the Brouwer fixed point theorem. Now, *if the points $\{P_i\}$ have been chosen sufficiently close together,* when the vector $V(P)$ passes through the vertical in a counterclockwise direction an edge labeled AB will appear, and when the vector passes through the vertical going clockwise an edge labeled BA will appear. The winding number can therefore be computed just like the index in §7. The cases where the vector reaches the vertical and then retreats are also counted properly this way: they will correspond to a sequence of points labeled AAA or BAB and will contribute zero to the index. To illustrate, Figure 8.2 repeats the curves of Figure 8.1 with the winding numbers worked out in this way. For convenience the subdivision points are chosen to be the same as the points at which vectors were drawn in Figure 8.1.

The index could now be adopted as the definition of the winding number except for the problem of deciding how close together the points of the subdivision $\{P_i\}$ should be in order that no circuit of the vector $V(P)$ be missed. In order to attack this problem, let us say that a subdivision $\{P_i\}$ of γ is ε-**dense** if any point inserted between two points of the subdivision will be within a distance ε of the points on either side (see Figure 8.3). Intuitively it is clear that the more dense the subdivision (i.e., the smaller the ε), the more certain that the index will equal the winding number. The question is, How dense a subdivision is needed? An answer of a sort is given in the following theorem.

Theorem

Let the continuous vector field V be defined on the closed path γ, and assume that V is never zero on γ. For any subdivision $\mathscr{P} = \{P_i\}$ of γ, let $I(\mathscr{P})$ be the index of the polygon $\mathscr{P}$ labeled according to the direction of the vectors V at the vertexes of $\mathscr{P}$. Then there exists a constant $\varepsilon > 0$ such that if $\mathscr{P}$ and $\mathscr{Q} = \{Q_j\}$ are any two ε-dense subdivisions of γ, then $I(\mathscr{P}) = I(\mathscr{Q})$.

In other words, the indexes obtained from the subdivisions of γ of ε-density or finer all agree. The number upon which they all agree must be the winding number. This leads to the following definition.

Definition

*Given a closed path γ and a continuous vector field V that is never zero on γ, the **winding number of V on γ**, notated $W(\gamma)$, is the index of the labeled polygon*

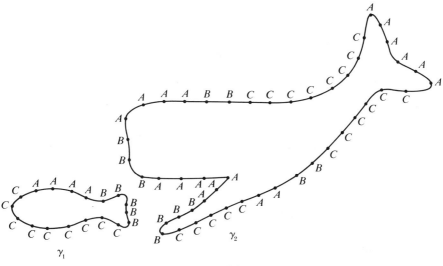

Figure 8.2

obtained from any ε-dense subdivision of γ, where ε is the constant supplied by the previous theorem.

This definition deserves some comment. We have now given three definitions of winding number, each more elaborate and more precise than its predecessor. The first definition, which involves watching the vector $V(P)$ while the point P traverses $γ$, is the intuitive essence of winding numbers but is difficult to apply. The second definition, involving watching only one direction, is the easiest to apply in examples and will be used to actually compute winding numbers in preference to the other definitions (see the example below). The third definition, which is clearly equivalent to the second, is the most combinatorial of the definitions and therefore the most appropriate for theoretical use. Thus it has been adopted officially and will appear consistently in the theory ahead. It can be used even if the vector field oscillates infinitely often about due north; all that is necessary is the continuity of the vector field. (It is also important to point out that the winding number is independent of the chosen direction and the labeling conventions used in computing the index. Although this is intuitively obvious, a proof is quite technical. A simple proof will be given in Chapter Six.) The one disadvantage of our official definition of winding numbers is that it is next to impossible to apply in examples because the theorem upon which it is based only half answers the question, How dense should the subdivisions be? The theorem only asserts that there is some measure ε of density beyond which all subdivisions have the same index. But the proof, to which we now turn, gives no means of finding this ε.

To begin the proof, recall the topological lemma of §6. This applies to γ because closed paths are compact. Therefore there is a constant $\varepsilon > 0$ such that every complete triangle with vertexes in γ has a side greater than ε. This ε will be the constant ε referred to in the present theorem. Let $\mathscr{P}$ be an ε-dense subdivision of γ. First examine what happens when one point is added to $\mathscr{P}$; say the point Q is inserted between the points P_i and P_{i+1} (Figure 8.3). The points P_i, Q, and P_{i+1} are the vertexes of a triangle, all of whose sides are less than ε. It follows that the labels for these vertexes cannot be a complete set. Now by examining all the possibilities one can verify that the insertion of Q cannot alter the index $I(\mathscr{P})$. For example, suppose the labels on P_i and P_{i+1} are AB. Then the label on Q must be an A or a B, and the index is unchanged by the insertion. If the labels read AA before insertion, the label on Q could be a B. Then an edge of type AB is added by the insertion of Q, but also an edge of type BA, so that the index is unchanged. What is specifically prevented by our choice of ε is that the labels read AC before insertion and ABC afterward. This would lead to an additional AB edge but also leads to a complete triangle, and so would contradict the topological lemma. The complete analysis of all possibilities is left to you.

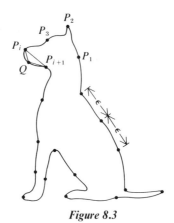

Figure 8.3

Now consider two ε-dense subdivisions $\mathscr{P} = \{P_i\}$ and $\mathscr{Q} = \{Q_j\}$. Starting with $\mathscr{P}$, we add the points of $\mathscr{Q}$ one at a time to obtain a new subdivision $\mathscr{R}$ consisting of the points of $\mathscr{P}$ and $\mathscr{Q}$ together. Since at each step the index is unaltered, $I(\mathscr{P}) = I(\mathscr{R})$. Similarly, $I(\mathscr{Q}) = I(\mathscr{R})$, and this completes the proof.

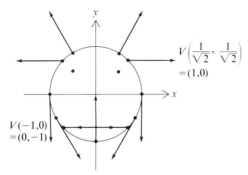

Figure 8.4

Example

Let us find the winding number of the vector field $V(x, y) = (2xy, y^2 - x^2)$ on the unit circle $x^2 + y^2 = 1$. First by means of a rough sketch (Figure 8.4) and the first definition of winding number, we conclude that $W(\gamma) = 2$. Next applying the second definition, we find the points on the circle where the vector V points north by solving simultaneously the equations

$$x^2 + y^2 = 1 \qquad \text{the circle}$$

$$\left. \begin{array}{r} 2xy = 0 \\ y^2 - x^2 \geq 0 \end{array} \right\} \quad \text{the vector } V \text{ points north}$$

The solutions are the points $(0, 1)$ and $(0, -1)$. At these points the x-coordinate of V tells whether the vector is moving counterclockwise or clockwise. In each case $2xy$ changes sign from $+$ to $-$, indicating that V moves from the first to the second quadrant or from direction A to direction B. Therefore each solution counts $+1$ as part of the winding number, and so we confirm that $W(\gamma) = +2$.

Exercises

1. Compute the winding number of $V(x, y) = (y, 1 - x^2)$ on the following curves.

(a) $x^2 + y^2 = 2x$ (b) $x^2 + y^2 = -2x$

(c) $x^2 + y^2 = 2y$ (d) $x^2 + y^2 = -2y$

2. Compute the winding number of $V(x, y) = (y(x^2 - 1), x(y^2 - 1))$ on the following curves.

(a) $x^2 + y^2 - 2x - 2y + 1 = 0$ (b) $x^2 + y^2 + x + y = \frac{1}{2}$

(c) $x^2 + y^2 = 1$ (d) $x^2 + y^2 = 4$

The official definition of winding number, combined with the index lemma, yields the following theorem.

The Fundamental Theorem on Winding Numbers

Let $\mathscr{D}$ be a cell with the closed path γ as boundary. If the continuous vector field V is never zero on $\mathscr{D}$, then $W(\gamma) = 0$.

In other words, if $W(\gamma) \neq 0$, then V has a critical point in $\mathscr{D}$. This theorem (due to Poincaré) is a first hint that the winding number can give information on the vector field *inside* the path. In the following sections many applications of this idea will appear. For the proof we apply the topological lemma to $\mathscr{D}$. Since V is never zero on $\mathscr{D}$, there is an $\varepsilon > 0$ such that every complete triangle with vertexes in $\mathscr{D}$ has a side of length greater than ε. Choose an ε-dense subdivision of γ. By definition, the index of this subdivision is the winding number of V on γ. Incorporating this subdivision into a triangulation of $\mathscr{D}$, by the index lemma the winding number equals the content of this triangulation. Now if enough vertexes are added to the triangulation so that the distance between adjacent vertexes is always less than ε, the triangulation will have no complete triangles. Then the content is zero, and this proves the theorem.

§9 ISOLATED CRITICAL POINTS

Let V be a continuous vector field and P a point. We suppose that V is defined not only at P but in some neighborhood of P. If P is an ordinary point, then $V(P)$ is not zero and it is possible to choose the neighborhood so that V is never zero there. If P is a critical point, then it may still be possible to choose a neighborhood so that V vanishes only at P in the neighborhood. Then P is called an **isolated critical point**. For example, all the critical points in Figure 7.1 are isolated.

Winding numbers become a tool for the study of isolated critical points in the following way: choose a circle γ about the critical point P so that within

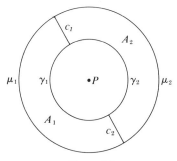

Figure 9.1

and on γ the vector field V never vanishes except at P. The **index of V at P**, denoted $I(P)$, is defined as the winding number $W(\gamma)$ of V on γ. Clearly there are many circles of this description about P. What makes the index useful is that it does not depend on which circle is chosen. Supposing that γ and μ are two such circles (see Figure 9.1), we will prove that $W(\gamma) = W(\mu)$. On the annulus A between the two circles, V is never zero. Adding two paths c_1 and c_2 cutting across the annulus, we divide the annulus and the two circles into halves. Let the halves of the two circles be called γ_1 and γ_2 for γ, and μ_1 and μ_2 for μ. These names are chosen so that if A_1 and A_2 are the two halves of the annulus A, then the boundary of A_1 consists of γ_1, μ_1, and the two cuts. while the boundary of A_2 consists of γ_2, μ_2, and the two cuts. By the fundamental theorem on winding numbers, the winding numbers of V on these boundary curves are zero. In the sum of these two winding numbers the contributions from the two cuts c_1 and c_2 cancel out because the edges labeled AB along them are counted both plus one and minus one. All that remains are the contributions from the parts of the two circles. The halves of the outside circle are taken counterclockwise as parts of the boundary of A_1 and A_2. Therefore the sum of the numbers of AB edges on γ_1 and γ_2 equals $W(\gamma)$; however, the parts of the inner circle are taken clockwise as parts of the boundary of A_1 and A_2, and therefore their contributions amount to minus $W(\mu)$. Thus $W(\gamma) - W(\mu) = 0$.

Since the winding number $W(\gamma)$ is the same no matter how small the circle γ, it follows that the index measures something about the vector field V that depends only on the behavior of V in an arbitrarily small area around P. Figure 9.2 gives some examples of critical points. In each case we give a picture of the integral paths in the vicinity of the point and a picture of the behavior of the vector field on a circle drawn around the critical point. The index is only a crude means of classifying critical points, since points of widely different types may have the same index. Nonetheless, the index turns out to be an extremely useful theoretical tool.

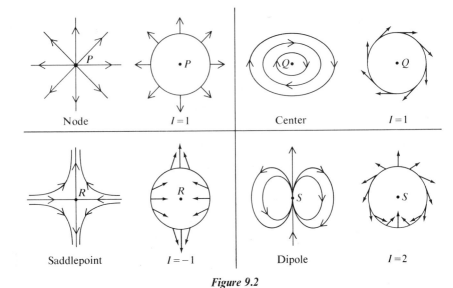

Figure 9.2

Exercises

1. Find the indexes for the critical points of the vector fields in Exercise 1 of §7.
2. Find the indexes for the following critical points.

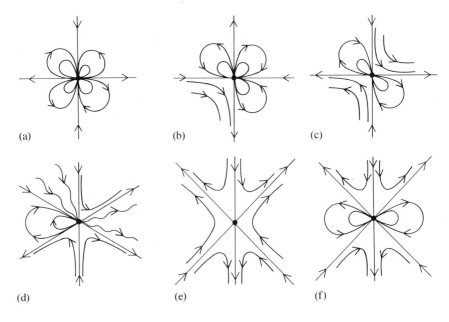

3. Use the result of Exercise 5 of §7 plus the topological lemma to give another proof that the index of a critical point does not depend on the choice of the circle around P.

We now give a survey of the different types of isolated critical points. First there is the **center**, which we introduced earlier, and a close relative the **focus** (Figure 9.3). The integral paths of a focus never reach the critical point but spiral endlessly about it. Foci may be stable or unstable.

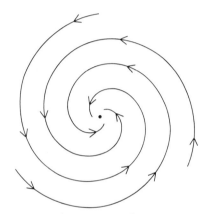

Figure 9.3   A focus.

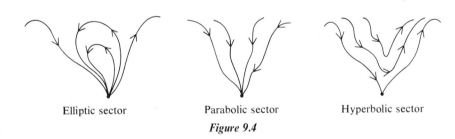

Elliptic sector   Parabolic sector   Hyperbolic sector

Figure 9.4

All other critical points, although we shall not prove this, are made up of sectors of the three types shown in Figure 9.4: **elliptic sectors**, where all paths begin and end at the critical point; **parabolic sectors**, where just one end of the path is at the critical point; and **hyperbolic sectors**, where the paths do not reach the critical point at all. A typical critical point might have sectors of all three types (Figure 9.5). The paths that divide each sector from the next are called **separatrixes**. It may happen that a critical point has only one type of

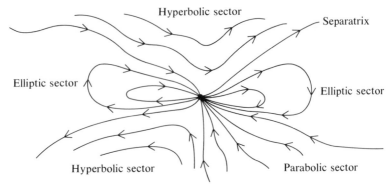

Figure 9.5   A typical isolated critical point.

sector. Those with only parabolic sectors are called **nodes**. Since a parabolic sector can be stable or unstable, nodes can be stable or unstable. The other types of sectors are unstable so the only stable critical points are the stable focus, node, and center. A critical point with only elliptic sectors is called a **rose**. An example is the dipole (Figure 9.2). A critical point with only hyperbolic sectors is called a **cross point**. Saddle points are cross points with four sectors. Still more complicated types are possible with an infinite number of sectors, and there may be nonisolated critical points. We shall always assume that critical points are isolated and have a finite number of sectors.

Exercises

4. Verify that the system of differential equations corresponding to the vector field $V(x, y) = (x + y, -x + y)$ has solution $x = Ke^t \sin(t)$, $y = Ke^t \cos(t)$. What type of critical point does V have at the origin? Is it stable or unstable?

5. Analyze the examples of critical points in Exercise 2 into sectors of different types.

6. The number of elliptic sectors plus the number of hyperbolic sectors is always even. Verify this for the examples of critical points illustrated above, and then prove it.

Let P be an isolated critical point other than a center or a focus. The object of the next few exercises is to derive a formula for the index of V at P in terms of the numbers e, p, and h of elliptic, parabolic, and hyperbolic

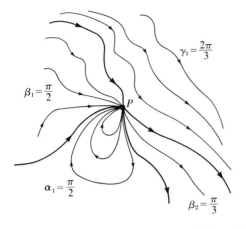

$\gamma_1 = \dfrac{2\pi}{3}$

$\beta_1 = \dfrac{\pi}{2}$

P

$\alpha_1 = \dfrac{\pi}{2}$

$\beta_2 = \dfrac{\pi}{3}$

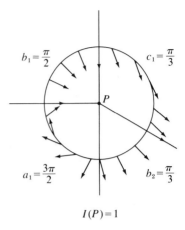

$b_1 = \dfrac{\pi}{2}$

$c_1 = \dfrac{\pi}{3}$

P

$a_1 = \dfrac{3\pi}{2}$

$b_2 = \dfrac{\pi}{3}$

$I(P) = 1$

Figure 9.6

sectors, respectively. Assume for simplicity that the sectors are bounded by separatrixes that enter P from a definite direction (in other words, have a tangent at P). Then each sector can be given an angle measure, namely the angle between the tangents to the separatrixes of that sector. In the example illustrated in Figure 9.6 there are four sectors: two of parabolic type and one each of elliptic and hyperbolic type. Thus $e = 1$, $p = 2$, and $h = 1$.

In general let $\alpha_1, \alpha_2, \ldots, \alpha_e$ be the angle measures of the elliptic sectors, $\beta_1, \beta_2, \ldots, \beta_p$ the angle measures of the parabolic sectors, and $\gamma_1, \gamma_2, \ldots, \gamma_h$ the angle measures of the hyperbolic sectors.

Exercises

7. Show that

$$\alpha_1 + \alpha_2 + \cdots + \alpha_e + \beta_1 + \beta_2 + \cdots + \beta_p + \gamma_1 + \gamma_2 + \cdots + \gamma_h = 2\pi$$

8. To compute the index, draw a circle about P and let the winding number of V be computed on this circle. Let $a_1, a_2, \ldots, a_e, b_1, b_2, \ldots, b_p, c_1, c_2, \ldots, c_h$ be the angles through which the vector V revolves in each of the corresponding sectors about P. Figure 9.6 gives an example. Explain why

$$a_i = \alpha_i + \pi \qquad i = 1, 2, \ldots, e$$
$$b_j = \beta_j \qquad j = 1, 2, \ldots, p$$
$$c_k = \gamma_k - \pi \qquad k = 1, 2, \ldots, h$$

9. Show that

$$I(P) = 1 + \left(\frac{e - h}{2}\right)$$

This is the desired formula. Note that in the end the number of parabolic sectors is irrelevant. Does this formula agree with previous computations of indexes?

10. In what way is an ordinary point like an isolated critical point with just two hyperbolic sectors? In what way unlike?

11. Among critical points of 4 and fewer sectors, find all the topologically distinct ones of index 3, 2, and 1.

---

§10 THE POINCARÉ INDEX THEOREM

There is an amazing connection between the indexes of the critical points and the winding numbers of the vector field.

The Poincaré Index Theorem

Let V be a continuous vector field. Let $\mathscr{D}$ be a cell and γ its boundary. Supposing that V is not zero on γ, then

$$W(\gamma) = I(P_1) + I(P_2) + \cdots + I(P_n)$$

where $P_1, P_2, \ldots, P_n$ are the critical points of V inside $\mathscr{D}$.

This is a remarkable theorem because it connects the behavior of the vector field inside the cell with its behavior on the boundary. For the proof we first point out that V can have only finitely many critical points in $\mathscr{D}$. If V had an infinite set A of critical points, then by compactness there would be a point P of $\mathscr{D}$ near A. By continuity $V(P)$ would be zero, so that P would be a nonisolated critical point. Therefore V can have only a finite number of critical points in $\mathscr{D}$. Let these points be $P_1, P_2, \ldots, P_n$. Around each we construct a circle γ_i that encloses P_i and no other critical point (see Figure 10.1). Then drawing paths between γ and the circles γ_i, and among the circles γ_i, we divide the region in $\mathscr{D}$ outside the circles into a number of cells D_k to each of which we may apply the fundamental theorem on winding

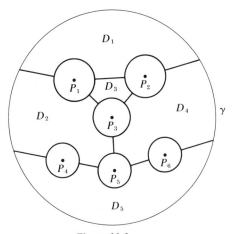

Figure 10.1

numbers. Then, just as with Figure 9.1, we find that

$$W(\gamma) = W(\gamma_1) + W(\gamma_2) + \cdots + W(\gamma_n)$$
$$= I(P_1) + I(P_2) + \cdots + I(P_n)$$

EXAMPLES OF PHASE PORTRAITS

Let us now examine a few specific systems of differential equations. Using only a crude sketching technique, one can still investigate a number of interesting phase portraits.

Consider the vector field $V(x, y) = (F, G) = (2xy, y^2 - x^2 - k^2)$, where k is a constant. There are two critical points at $(0, k)$ and $(0, -k)$, $k \neq 0$. They are found by solving simultaneously the equations $F = 2xy = 0$ and $G = y^2 - x^2 - k^2 = 0$. To sketch the phase portrait it suffices to draw the vector field along the curves given *separately* by the equations $F = 0$ and $G = 0$. These curves, called **critical curves**, intersect at the critical points. In this case the critical curves are the coordinate axes ($F = 0$) and a hyperbola ($G = 0$). Along the critical curves the vectors are either horizontal or vertical, and therefore particularly easy to draw. This has been done in Figure 10.2 (thick vectors). In between the critical curves there can be no vertical or horizontal vectors. Therefore in these regions determined by the critical curves *all the vectors point in the direction of just one quadrant*. The vector field can thus be filled in with roughly parallel vectors in these regions (thin vectors). The magnitude of the vectors is not important, since we are in-

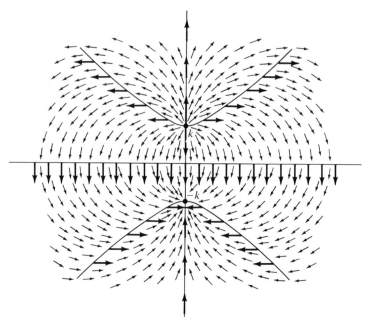

Figure 10.2   The vector field V $(k \neq 0)$.

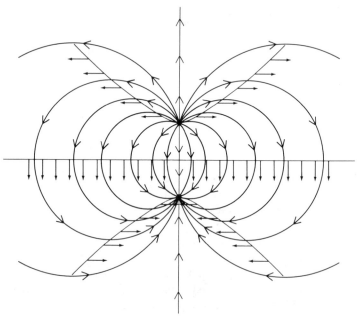

Figure 10.3   The phase portrait of V $(k \neq 0)$.

terested in them only as the tangents of the integral curves. They can be drawn all of the same length. The integral paths are now easily sketched in (Figure 10.3). We find that the critical points are nodes, one stable and one unstable. The phase portrait represents a flow from one critical point to the other.

Now suppose the constant k tends to zero. Then the two critical points move toward each other, meeting at the origin in the single critical point of the vector field $V(x, y) = (2xy, y^2 - x^2)$. The critical curves now include, in addition to the axes as before, the set of $45°$ lines, which come from the hyperbola. Using the same technique as before (Figures 10.4 and 10.5), we find that the critical point is a dipole. Thus the two nodes of the original

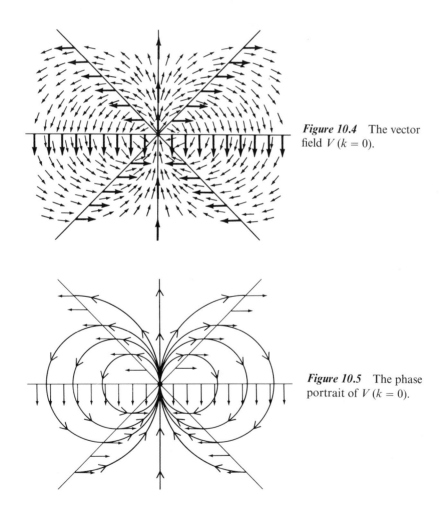

Figure 10.4   The vector field V $(k = 0)$.

Figure 10.5   The phase portrait of V $(k = 0)$.

vector field have joined to form a single critical point of index two! This is expressed by saying that the dipole is the **confluence** of the two nodes. You are urged to sketch these and other phase portraits personally, because no description can replace this experience.

THE VOLTERRA PREY-PREDATOR EQUATIONS

These concern the growth of two species, one the prey of the other—say foxes and rabbits. Supposing that the rabbits have abundant forage, we expect the rabbit population to grow at a rate simply proportional to the size of the population (exponential growth). In other words, if $x(t)$ denotes the number of rabbits in the population at time t, then $x' = ax$, where the constant a represents the difference between the birth and death rates for rabbits. But what about the foxes? Supposing that some proportion of the encounters between foxes and rabbits results in the death of a rabbit, the foxes represent a check on the growth of the rabbit population. Assuming that the product xy, where $y(t)$ is the number of foxes at time t, is a good measure of the number of encounters between foxes and rabbits, then the differential equation for rabbits becomes

$$x' = ax - bxy$$

where the constant b reflects the proportion of fox-rabbit encounters that end badly for the rabbit. On the other hand, the fox population will be governed by the equation

$$y' = -cy + dxy$$

The coefficient of y is the natural growth rate of the foxes ignoring the rabbits, by analogy with a. This is negative, reflecting the fact that the foxes depend on the rabbits for prey. The last term is positive, since encounters with rabbits are good for foxes.

Thus we are led to consider the vector field $U(x, y) = (ax - bxy, -cy + dxy)$. This vector field and the corresponding phase portrait are sketched in Figures 10.6 and 10.7. There are two critical points at $(0, 0)$ and $(c/d, a/b)$. The critical curves are the axes and the lines $x = c/d$ and $y = a/b$. The critical points turn out to be a center and a saddle point. You may well wonder how it was decided that the point $(c/d, a/b)$ was a center, since the sketched data must seem equally consistent with a focus. In this case there is luckily a clever technique for graphing the integral paths in the first quadrant (see Exercise 4) that reveals them as closed paths. In general, however, this is one of the most difficult problems in the qualitative theory of differential equations: identifying closed integral paths when they exist. Closed paths are important because they correspond to periodic solutions of the system of differential

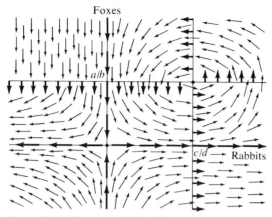

Figure 10.6 The vector field of the prey-predator equations.

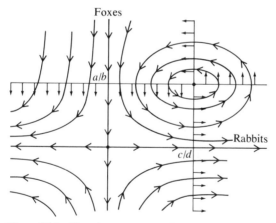

Figure 10.7 The phase portrait of the prey-predator equations.

equations. In the next section we encounter a necessary condition for the existence of closed paths (Exercise 2 in that section), but no really comprehensive sufficient condition is known.

From the point of view of the foxes and rabbits, only the first quadrant portion of the phase portrait is of interest. The closed paths correspond to a cyclical growth and decay pattern for both fox and rabbit populations. The two cycles are out of phase with each other, as in Figure 10.8.

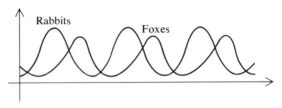

Figure 10.8

Exercises

1. Examine the results of Exercise 1 of §8. Do you find any evidence for the Poincaré index theorem?

2. Sketch the phase portraits of the following vector fields without solving the corresponding differential equations (* denotes a center, ** a focus).

*(a) $V(x, y) = (y, x^2 - 1)$ (b) $V(x, y) = (y, x^2)$

*(c) $V(x, y) = (y, x - x^3)$ **(d) $V(x, y) = (y - x, xy - 2x)$

(e) $V(x, y) = (x^2 - y, x + 3y)$ **(f) $V(x, y) = (x^2 - y, xy - x)$

(g) $V(x, y) = (y^2 - x, xy - y)$ (h) $V(x, y) = (x(y^2 - 1), y(x^2 - 1))$

3. Using parts (a) and (b) of the previous problem, discuss the confluence of a saddle point and a center.

4. By solving equation (2) of §7, show that the integral paths of the prey-predator equations are given by the equation

$$y^a e^{-by} = Kx^{-c}e^{dx} \qquad K \text{ constant}$$

This equation cannot be solved for y in terms of x but can be graphed by the following trick, due to Volterra. Introduce two new variables z and w, related to x and y by the equations

$$z = y^a e^{-by} \qquad w = Kx^{-c}e^{dx}$$

These two equations can be graphed simultaneously on a pair of axes in which all four axes represent different variables. This is demonstrated in Figure 10.9. In the second quadrant the equation of z and y is plotted, in the fourth quadrant the equation of w and x is plotted, while the line $z = w$ is plotted in the third quadrant. Choosing a point on the line $z = w$ and following around to the first quadrant will produce points on the integral path. Use this scheme to graph the whole path. Show how different values of K yield concentric integral paths.

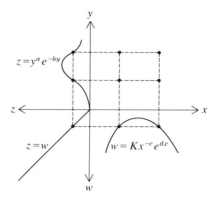

Figure 10.9

5. Consider the following prey-predator relationship. The prey is the gypsy moth caterpillar, and the predator is one of several parasitic wasp larvae that attack the caterpillars. A spray program is to be instituted to control the caterpillars, but the spray kills wasp and caterpillar alike. Suppose first that it is a question of a single spraying and the result is the death of an equal proportion of the wasp larvae and caterpillar populations. Explain how the timing of the spraying is crucial if this is not to lead to an eventual increase in the number of caterpillars. Next suppose that a systematic program of repeated sprayings is instituted, resulting in an increase in the death rate for wasp larvae and caterpillars. Explain how this could also lead to an eventual increase in the number of caterpillars.

6. Develop an interaction theory of the prey-predator type for two species, each of which has a positive natural growth rate, but which compete with each other? What are some possible fields of application for this theory? What other hypothetical relationships between two species can you discuss?

§11 CLOSED INTEGRAL PATHS

Among the closed paths to which one might apply the Poincaré index theorem, none are more important than the closed integral paths, which represent periodic solutions of the corresponding system of differential equations. The winding number of the vector field on such curves is given by another theorem of Poincaré.

Theorem

The winding number of a vector field on a closed integral path is one.

The following proof is due to H. Hopf. Let γ be a closed integral path for the continuous vector field V. Pick a point P on γ. For every other point S on γ, let s be the arc length from P to S along the curve γ, taken in the direction indicated by the vectors V. Given the base point P, which will be fixed for the rest of the proof, the arc length s uniquely determines the point S. If L is the total length of γ, then for every number s, $0 < s < L$, there is exactly one point S whose distance along γ from P is s (see Figure 11.1a).

Consider next a new plane with coordinate axes x and y. Within this plane, concentrate on the triangle ABC consisting of the points (x, y) such that $0 \leq x < y \leq L$ (see Figure 11.1b). To each point (x, y) in this triangle there corresponds a pair of points X and Y on γ, such that the point Y is further along γ from P than X. The points on the hypotenuse (x, x) are exceptional and correspond only to a single point of γ. In addition, the corner point $(0, L)$ corresponds to a single point, the point P itself.

Using this correspondence, a vector field can be defined on the triangle ABC as follows: let $U(x, y)$ be the unit vector in the direction of the secant

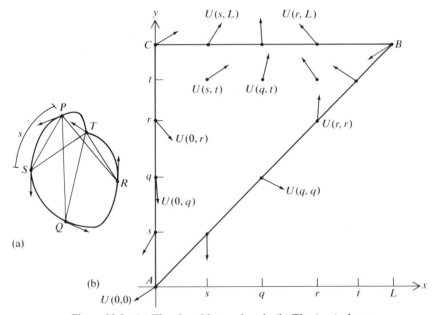

Figure 11.1   (a) The closed integral path. (b) The (x, y) plane.

vector on γ from X to Y. Figure 11.1 gives a few examples. Each secant drawn on Figure 11.1a corresponds to a vector of U drawn on Figure 11.1b. The vector field U is clearly continuous. At the exceptional points (x, x), U is defined to be the unit tangent vector pointing in the same direction as $V(X)$. This makes U continuous on the whole triangle ABC, since these tangent vectors are the limits of the secant vectors of γ, which are the values of U inside the triangle. Similarly, for reasons of continuity, $U(0, L)$ must be defined as the unit vector pointing in the opposite direction from $V(P)$.

To complete the proof, we compute the winding number of U around the triangle ABC. This can be divided into two parts, the first arising from the hypotenuse, the second arising from the two legs of the triangle. On the hypotenuse, U simply repeats the vectors of V around γ, so the first part equals $W(\gamma)$. On the legs, U makes a single clockwise revolution (half a revolution on each leg). Therefore the total winding number of U is $W(\gamma) - 1$. On the other hand, since U is never zero inside or on the triangle, by the fundamental theorem of winding numbers, this winding number must be zero. This completes the proof.

APPLICATION TO CONTOUR LINES

A continuous real valued function $\phi(x, y)$ defined on a region $\mathscr{D}$ of the plane defines a surface in space: the graph of the equation $z = \phi(x, y)$. An example is given in Figure 11.2a together with some of the contour lines or

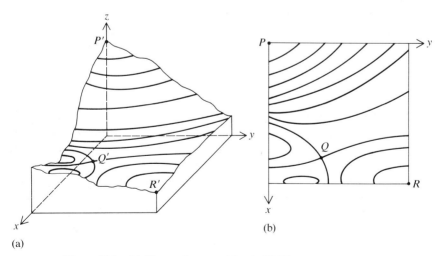

(a)

(b)

Figure 11.2   (a) The surface $z = \phi(x, y)$. (b) The contour map.

lines of constant height: solutions of the equation

$$\phi(x, y) = \text{a constant}$$

Figure 11.2b contains the corresponding contour map. Supposing ϕ differentiable, and taking the derivative of the above equation, we obtain

$$\frac{\partial \phi}{\partial x} dx + \frac{\partial \phi}{\partial y} dy = 0$$

Thus the contour lines satisfy the differential equation

$$\frac{dy}{dx} = \frac{-\partial \phi / \partial x}{\partial \phi / \partial y}$$

In other words, the contour lines are integral paths for the vector field

$$V(x, y) = \left(\frac{\partial \phi}{\partial y}, -\frac{\partial \phi}{\partial x} \right) \tag{1}$$

By studying the vector field V we gain information on contour lines and vice versa. For example, it is clear that different contour lines cannot intersect or even have points near each other. Therefore V cannot have nodes or in general any critical point with elliptic or parabolic sectors. The most common critical points are centers (which correspond to *peaks* such as P and P' in Figure 11.2, and *bottoms* such as R and R') and saddle points (which correspond to *cols* such as Q and Q'). Other cross points can also occur.

Example

Consider the vector field $V(x, y) = (1 - x^2, 2xy)$. By inspection (try it!) we see that $\phi(x, y) = y - x^2 y$ is a height function for V. There are two critical points, $(-1, 0)$ and $(+1, 0)$. At the critical points ϕ has the value zero, so that graphing the equation $\phi(x, y) = y - x^2 y = 0$ gives the separatrixes of the phase portrait. In this case these are the lines $y = 0$, $x = 1$, $x = -1$. The other integral paths are graphs of the equation $\phi(x, y) = y - x^2 y = k$, where k is any constant. The phase portrait is drawn in Figure 11.3. Both critical points are saddle points; the x-axis is a sort of ridge with two mountain ranges rising below it and one rising above it.

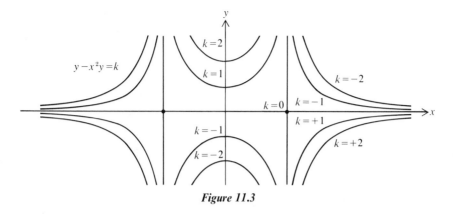

Figure 11.3

Exercises

1. In the proof of Poincaré's theorem on closed integral paths, explain why the winding number of U on the two legs of ABC is minus one.

2. Prove that every closed integral path encloses a critical point. This is an important result. It implies that the search for closed integral paths must take place around the critical points of a vector field.

3. The following vector fields V have height functions ϕ with which they are related by equation (1). Find these height functions and use them to sketch the phase portraits.

(a) $V(x, y) = (y, -x)$ (b) $V(x, y) = (x, -y)$

(c) $V(x, y) = (x + y, -y)$ (d) $V(x, y) = (-2xy, y^2 - x^2 - 1)$

(e) $V(x, y) = (y^2 - x^2 - 1, 2xy)$ (f) $V(x, y) = (x^3 - 3xy^2, y^3 - 3yx^2)$

4. Let ϕ be the height function on a desert island. Let V be the corresponding vector field. Prove that the winding number of V along the shoreline is 1. Then supposing that the only critical points are peaks, bottoms, and cols, prove that $P - C + B = 1$, where P is the number of peaks, C is the number of cols, and B is the number of bottoms. Of what other combinatorial result does this remind you?

5. Referring to the previous exercise, suppose that the island is no longer desert but has L lakes. Prove that $P - C + B = 1 - L$. Verify this by carefully identifying and counting all the critical points on the island in Figure 11.4.

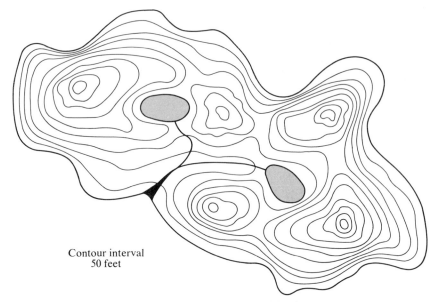

Figure 11.4   A deserted island.

6. Cross points other than saddle points may occur in topography. Figure 11.5 shows a **monkey saddle**. Explain the derivation of this name, and draw a contour map of this surface. What is the index of the critical point? Modify the results of the two preceeding exercises to apply to surfaces with monkey saddles.

Figure 11.5   A monkey saddle.

7. Draw contour lines on the map in Figure 11.6. The contour interval should be 200 feet. Use the elevations provided, and mark all the critical points.

Figure 11.6   Treasure Island.

8. Draw contour maps of islands with the following specifications:

(a) one lake and one peak

(b) two saddle points and two lakes

(c) one saddle point and one monkey saddle

§12 FURTHER RESULTS AND APPLICATIONS

DUAL VECTOR FIELDS

Given the vector field $V(x, y) = (F(x, y), G(x, y))$, the **dual vector field** is defined by

$$V^*(x, y) = (-G(x, y), F(x, y))$$

Exercises

1. Search among previous examples and exercises for pairs of vector fields and their duals. There are at least four such pairs. Draw their phase portraits together, and study the relationship between them.

2. Show that $V^*(P)$ is the vector $V(P)$ rotated $90°$ counterclockwise. Use this fact to explain that

(a) V and V^* have the same winding numbers on all closed paths

(b) V and V^* have the same critical points with the same indexes

(c) the integral paths of V^* are the **orthogonal trajectories** of the integral paths of V; that is, the two families of curves always meet at right angles

3. Conduct a study of the behavior of critical points under dualization. This can be done systematically for complicated critical points by sketching the orthogonal trajectories sector by sector. Start, however, by simply examining the examples of dualization mentioned in Exercise 1. Is it possible to predict the number of sectors of each type in the dual critical point from these numbers for the original critical point? Is there any connection between stability of a critical point for a given field and for its dual?

4. The dual of the confluence of two critical points is the confluence of their duals. Use this principle plus the examples in §10 of the confluence of two nodes and the confluence of a center and a saddle point to discuss the confluence of two centers and the confluence of a node and a saddle point. Also determine the confluence of two saddle points (use Figure 11.3).

5. What is the relation of the double dual V^{**} to the original vector field V?

GRADIENTS

Consider a real valued function $\phi(x, y)$ defined on a region of the plane. In §11 we considered the associated vector field

$$V = \left(\frac{\partial \phi}{\partial y}, -\frac{\partial \phi}{\partial x} \right)$$

whose integral paths are the contour lines of ϕ. The dual

$$V^* = \nabla \phi = \left(\frac{\partial \phi}{\partial x}, \frac{\partial \phi}{\partial y} \right)$$

is equally important and is called the **gradient** of ϕ. In this context ϕ is called a potential function for V^*. The integral curves of V^* are called **lines of steepest ascent**, because the vector V^* always points in the direction of steepest ascent on the surface defined by ϕ, in other words, in the direction of greatest increase of ϕ.

Exercises

6. Draw the lines of steepest ascent for the potentials of Exercise 3 of §11. Do the same for the map in Figure 11.4. Don't forget that water runs down hill by a path of steepest descent.

7. What types of critical point are possible for gradient vector fields?

8. Barometric pressure is a potential function that plays a crucial role in determining the world's weather. The integral paths of V in this case are called **isobars**. Neglecting the effect of the earth's rotation (the Coriolis effect), the integral paths of V^* are the **streamlines** of the wind, which moves from high to low pressure. Fill in the isobars and streamlines on the weather map in Figure 12.1. Use a contour interval of 0.1, and remember that the prevailing winds especially over water (i.e., at the edges of the map) are from the west at these latitudes. Find the winding number of V^* around the edge of the map. Label all critical points, and verify that Poincaré's index theorem is satisfied. (Actually, the Coriolis effect produces a flow of air *parallel* to the isobars, the so-called **geostrophic flow**.)

9. Much of the terminology of gradients comes from electrostatics, where potential energy is in fact the potential function. The integral paths of V are called **equipotential lines**, and the integral paths of V^* are the **field lines**

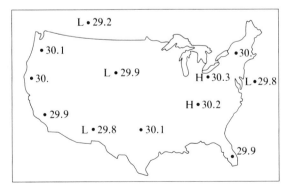

Figure 12.1 Data from 1400 hours July 12, 1974.

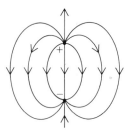

Figure 12.2

along which the electrostatic force acts. Figure 12.2 shows the field lines produced by a pair of opposite charges. Topologically this is equivalent to the phase portrait of Figure 10.3, although there the integral paths are circles. This field is called the **dipole field**. Imagine viewing this field from a great distance. The two critical points would flow together and appear like a single critical point: the dipole.

Draw the **quadrupole field**, the field associated with the following arrangement of four charges. There will be a fifth critical point in the center where the charges balance.

$$
\begin{array}{cc}
\overset{-}{\bullet} & \overset{+}{\bullet} \\[2em]
\overset{+}{\bullet} & \overset{-}{\bullet}
\end{array}
$$

What single critical point will the quadrupole resemble when seen from a great distance? Draw the field associated with a crystalline substance composed of dipoles as shown below. This is called an **electret**.

$$
\begin{array}{cccc}
\overset{+}{\bullet} & \overset{-}{\bullet} & \overset{+}{\bullet} & \overset{-}{\bullet} \\[1.5em]
\overset{-}{\bullet} & \overset{+}{\bullet} & \overset{-}{\bullet} & \overset{+}{\bullet} \\[1.5em]
\overset{+}{\bullet} & \overset{-}{\bullet} & \overset{+}{\bullet} & \overset{-}{\bullet} \\[1.5em]
\overset{-}{\bullet} & \overset{+}{\bullet} & \overset{-}{\bullet} & \overset{+}{\bullet}
\end{array}
$$

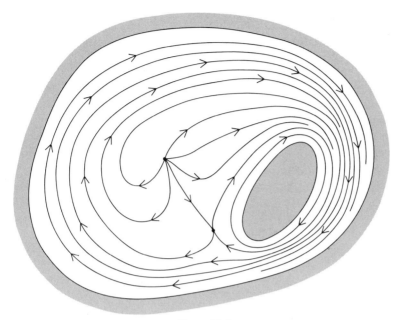

Figure 12.3

FLUID FLOW

Consider a spring-fed lake as in Figure 12.3. The velocity vectors for the water (ignoring vertical motion) form a continuous vector field. Fluid flows like this provide some of the most interesting and important examples of vector fields. They do not generally arise from potential functions. In Figure 12.3 the integral paths, called **streamlines**, spiral out toward the shore, which on this account is called a **limit cycle**.

Exercises

10. Describe the nature of the critical points associated with the following common phenomena:

(a) springs
(b) sinks (subterranean exits for the water)
(c) eddies

Cross points in this context are called **stagnation points**. Are other types of critical points possible?

11. Let S_1 be the number of springs, S_2 the number of sinks, S_3 the number of saddle points, and I the number of islands. Assuming that these are the only critical points, show that

$$S_1 + S_2 - S_3 = 1 - I$$

12. Draw streamlines for lakes with the following features:

(a) one island, one spring, and one eddy

(b) a spring and two eddies

(c) a spring, a sink, and one stagnation point

(d) two islands and a spring

(e) one island and *no* critical points

Notes. The qualitative theory of differential equations is a large branch of mathematics that uses a wide variety of techniques. If you are interested in further study of systems of differential equations, you should consult the elementary introduction in Simmons' book [28] and then look at the treatises by Lefschetz [22], Nemytskii and Stepanov [24], and Hirsch and Smale [13]. This book has drawn material from all these works.

three

Plane Homology and the Jordan Curve Theorem

§13 POLYGONAL CHAINS

The goal of this chapter is to prove a famous theorem, first stated by Jordan in 1887, to the effect that every closed path (Jordan curve) divides the plane into two pieces, an inside and an outside. At first glance this may seem trivial, but its simple statement is misleading. The difficulty lies with the possible complexity of Jordan curves. Even the second curve in Figure 13.1 does not

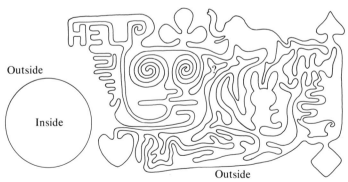

Figure 13.1   Two Jordan curves.

adequately convey this complexity, for this example is not only continuous but also differentiable, possessing a tangent at each point. It is possible to find examples of Jordan curves that are nowhere differentiable. For example, there is the **snowflake curve** (Figure 13.2), the invention of von Koch (1906). Such curves cannot be drawn exactly but only approximated. The construction of the curve begins with a bare equilateral triangle (Figure 13.2a). On each side of this triangle we place another triangle one third the size of the original and erase their common edge (Figure 13.2b). Repeating this operation indefinitely (Figures 13.2c, d, ...) we obtain in the limit a closed path that on account of its constant cornering is without tangents. Imagine for a moment a curve combining the features of Figures 13.1 and 13.2. This gives a better idea of the difficulty presented by an arbitrary Jordan curve.

Our informal presentation of the theorem conceals other difficulties. What is the meaning of the terms "inside" and "outside"? What does it mean for one set to "divide" another into pieces? Luckily the notion of connectedness supplies the means for making these ideas precise. Here is a formal statement of the theorem.

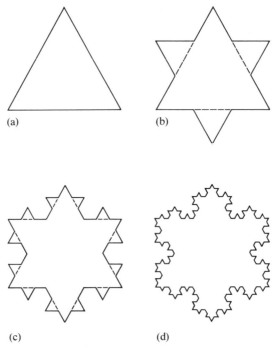

Figure 13.2   Approximations to the snowflake curve.

The Jordan Curve Theorem

Let $\mathscr{J}$ be a Jordan curve. Then the complement of $\mathscr{J}$ in the plane, $\mathscr{J}'$, is not connected but consists of two disjoint connected pieces, one of which is bounded (called the **inside***) and one of which is not bounded (called the* **outside***). The curve $\mathscr{J}$ forms the boundary for both pieces.*

Considering the deceptive simplicity of this statement, it is not surprising that the first proofs of the theorem to be offered (including Jordan's own) were incorrect. The first generally accepted proof was given in 1905 by Veblen. The proof given here is essentially the same following the treatment by Newman [25]. The proof depends on homology, a fundamental tool of combinatorial topology developed by Poincaré in a series of papers published from 1895 to 1904 growing out of his work on differential equations. The proof of Jordan's theorem provides a good opportunity to become familiar with Poincaré's ideas in the plane before studying them on more complicated surfaces. Homology occupies the bulk of the chapter, the proof of Jordan's theorem appearing in the last two sections. The remainder of this section is devoted to some preliminary ideas.

Definition

A **polygonal chain** *is a subset of the plane formed from a finite sequence of straight line segments parallel to the coordinate axes, each segment sharing endpoints with adjacent segments in the sequence.*

Polygonal chains are in many ways like paths. They differ from paths only in that they consist of straight line segments and they may cross themselves (see Figure 13.3). Like paths, polygonal chains have two endpoints. If these are the points P and Q, we say that P and Q are **connected** by the chain. Polygonal chains are important for two reasons. In the first place, as the union of a *finite* number of straight line segments, *polygonal chains can be treated combinatorially*. In fact, the homology developed in this chapter will be used largely

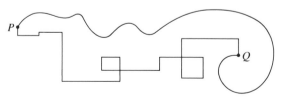

Figure 13.3   The points P and Q are connected by a path and a polygonal chain.

to manipulate these chains. In the second place, polygonal chains can be used to determine the connectedness or nonconnectedness of an open set, as the following theorem explains.

Polygonal Chain Theorem

Let G be an open set. Then G is connected if and only if every pair of points of G can be connected by a polygonal chain in G.

If every pair of points in G can be connected by a polygonal chain in G, then the connectedness of G follows immediately from Exercise 10 of §3 (whether G is open or not). Conversely, suppose G is connected. We shall prove that every pair of points in G can be connected by a polygonal chain in G. The proof is by contradiction. Suppose there is a pair of points P and Q that cannot be connected by a polygonal chain in G. Then the set G is divided into two sets, the set A consisting of those points that can be connected with P by a polygonal chain in G and the set B consisting of those points that cannot be connected with P in this way. These sets are not empty: $P \in A$, $Q \in B$. Therefore since G is connected, one of these sets contains a point near the other. Suppose that A contains a point R near B. Since G is open, R has a neighborhood entirely contained in G (see Figure 13.4). This neighborhood contains a point S of B because $R \leftarrow B$. Now R can be connected with S by a polygonal chain lying in the given neighborhood of R and hence lying in G, and it can also be connected with P by a polygonal chain lying in G. Therefore P can be connected with S by such a chain, contradicting the fact that S is in B. A similar contradiction results if instead B contains a point near A.   Q.E.D.

Consider the application of this theorem to the proof of the Jordan curve theorem. If $\mathscr{J}$ is a Jordan curve, then $\mathscr{J}$ is compact and hence closed, so that the complement $\mathscr{J}'$ is open and the preceeding theorem is applicable. To prove the Jordan curve theorem, we must establish that $\mathscr{J}'$ is not connected. This is accomplished by finding two points P and Q in $\mathscr{J}'$ that cannot be connected by a polygonal chain in $\mathscr{J}'$. Then to prove that $\mathscr{J}'$ consists of

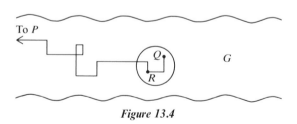

Figure 13.4

exactly two connected pieces, we will prove that any third point R can be connected to either P or Q in $\mathscr{J}'$.

Exercises

1. Let γ be a path and P a point of γ not an endpoint. Show that γ minus P is not connected but is divided into two connected subsets. This is a Jordan curve theorem in one dimension. What would be the statement of a Jordan curve theorem in three dimensions?

2. Find the perimeters and enclosed areas of the successive approximations to the snowflake curve. (Recall that an equilateral traingle of side s has area $s^2\sqrt{3/4}$.) By passage to the limit show that, although the snowflake curve encloses a finite area, it has infinite perimeter.

3. Another pathological curve is the **space filling curve** (Peano, 1890). Some approximations to such a curve are drawn in Figure 13.5. In the limit this curve passes through every point of the square. Space filling curves are not paths because they cross themselves. Use the result of Exercise 1 to prove that a path cannot be space filling. Show that every rectangle, no matter how small, contains a point not on a given path.

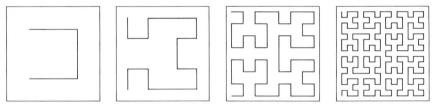

Figure 13.5

4. Let G be an open set and P a point of G. Define the set G_P to consist of all points of G that can be connected with P by a polygonal chain in G. Prove that G_P is connected and open. Show that G_P becomes disconnected with the addition of a single further point from G.

5. Let $\mathscr{J}$ be the unit circle. Use the intermediate value theorem to prove that any path beginning inside $\mathscr{J}$ and ending outside $\mathscr{J}$ (where inside and outside have their obvious meaning) must intersect $\mathscr{J}$. Use this result to give a complete proof of the Jordan curve theorem in this case.

6. Prove the Jordan curve theorem for the square with the equation $|x| + |y| = 1$.

§14 THE ALGEBRA OF CHAINS ON A GRATING

In order to study the polygonal chains defined in the last section, we introduce
the following notion: a **grating** $\mathscr{G}$ is a rectangular portion of the plane with
sides parallel to the coordinate axes and with a finite number of additional
lines drawn across the rectangle parallel to the sides. The intersections of all
these lines we call **vertexes**. These divide the lines into **edges**, while the edges
divide the rectangle into **faces**. In addition, it will prove extremely convenient
if we adopt the convention that the outside of the rectangle (in the obvious
sense) is an additional face. Thus in Figure 14.1 we have a grating with 63
vertexes, 110 edges, and 49 faces. (Check this!) Gratings will play a role in this
chapter analogous to the role that triangulations played in Chapter Two.

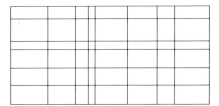

Figure 14.1   A grating.

Like triangulations, gratings are a type of complex. If we imagine a point at
infinity added to the plane, the grating may be regarded, by stereographic
projection, as if it were on a sphere. Then the outside face is a cell just like the
other faces. The theory of gratings applies equally to the sphere and the plane.
Note that any polygonal chain can be incorporated into a grating by first
extending the line segments of the chain until they meet the sides of the
rectangle (Figure 14.2). In this way gratings can be used to study polygonal
chains.

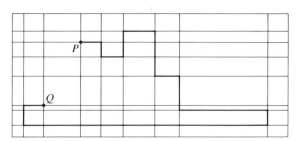

Figure 14.2   Incorporating a polygonal chain
in a grating.

The study of gratings is simplified by the adoption of a uniform terminology for faces, edges, and vertexes. All these basic counting blocks of combinatorial topology are to be called simplexes—**0-simplexes** for vertexes, **1-simplexes** for edges, and **2-simplexes** for faces. Frequently an argument or definition makes sense for all three types of simplex simultaneously. This terminology allows one statement to apply to all cases. Whenever possible the old terms face, vertex, and edge will be used to preserve the geometric flavor of the subject.

A union of simplexes is called a **chain**. Like simplexes, chains come in three types—a **0-chain** is a set of vertexes, a **1-chain** is a set of edges, and a **2-chain** is a set of faces. A chain is always of one of these types; there are no mixed chains. Figure 14.3 gives some examples. Clearly, polygonal chains are examples of 1-chains.

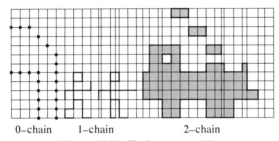

0–chain   1–chain   2–chain

Figure 14.3   Chains on a grating.

We now introduce an algebra of chains that is designed to handle the counting of chains for us. The counting, however, is of that peculiar sort where only the evenness or oddness of the result is used. You have seen the value of this type of counting already in Sperner's lemma. Actually we introduce three algebras, one for each type of chain. Suppose C_1 and C_2 are given k-chains, where k may be 0, 1, or 2. (We are using uniform terminology to make three definitions at once!) The sum $C_1 + C_2$ is defined to be the k-chain (same k!) made up of the k-simplexes in C_1 or C_2 *but not in both*. The addition sign $(+)$ is appropriate because this operation obeys all the usual algebraic laws of addition:

(a) *commutative law*: $C_1 + C_2 = C_2 + C_1$ for all k-chains C_1 and C_2.

(b) *associative law*: $C_1 + (C_2 + C_3) = (C_1 + C_2) + C_3$ for all k-chains C_1, C_2, and C_3.

(c) *zero*: There is a unique k-chain $\varnothing$ such that $C + \varnothing = C$ for all k-chains C.

(d) *inverses*: Each k-chain C has a unique inverse chain D such that $C + D = \varnothing$.

Examples of these operations of addition are given in Figure 14.4. An algebraic system satisfying just these four laws is called a (commutative) **group**.

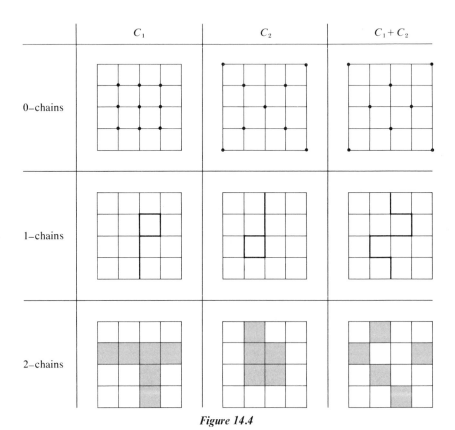

Figure 14.4

Thus each grating $\mathscr{G}$ determines three groups called the **chain groups** of $\mathscr{G}$. The proofs of these properties are left as an exercise. The most interesting to prove is the associative law. There one shows that both $(C_1 + C_2) + C_3$ and $C_1 + (C_2 + C_3)$ consist of those simplexes contained in *just one* or *all three* of the chains C_1, C_2, and C_3. More generally, the sum of n chains $C_1 + C_2 + \cdots + C_n$ turns out to consist of those simplexes contained in an *odd* number of the chains $C_1, C_2, \ldots, C_n$. The zero chain turns out to be the empty chain $\varnothing$ containing no simplexes.

Exercises

1. Given the k-chains C_1, C_2, and C_3, determine the k-chains $C_4 = C_1 + C_2$ and $C_5 = C_1 + C_2 + C_3$ in each of the following cases:

	C_1	C_2	C_3
$k = 0$			
$k = 1$			
$k = 2$			

2. Prove that addition of k-chains satisfies all the usual laws of addition. Also prove that the sum of n chains consists of those simplexes contained in an *odd* number of the summands.

3. Let C be a k-chain. Prove that C is compact if $k = 0$ or $k = 1$. Find an example of a noncompact 2-chain.

4. For each part of Exercise 1 solve the equation given below for the unknown k-chain C. The solutions should be given in two ways: by an illustration and by a formula expressing C in terms of C_1, C_2, and C_3.

(a) for $k = 0$, $C + C_1 = C_3$

(b) for $k = 1$, $C + C_5 = C_1$

(c) for $k = 2$, $C + C_4 + C_5 = \emptyset$

§15 THE BOUNDARY OPERATOR

The boundary operator, ∂, connects the algebra of k-chains with the algebra of $(k-1)$-chains on a grating: if C is a k-chain, the **chain boundary** $\partial(C)$ is a $(k-1)$-chain. Thus the boundary will only be defined for $k = 1$ or 2. In those cases we define $\partial(C)$ to be the chain of all $(k-1)$-simplexes that are contained in an *odd* number of the k-simplexes of C. Figure 15.1 gives some examples.

The most important property of the boundary operator is its **additivity**: if C_1 and C_2 are k-chains ($k = 1$ or 2), then

$$\partial(C_1 + C_2) = \partial(C_1) + \partial(C_2) \tag{1}$$

In words, the boundary of a sum is the sum of the boundaries. To prove this, let S be any $(k-1)$-simplex. Let n_1 and n_2 be the numbers of k-simplexes of C_1 and C_2, respectively, that contain S. Thus S is in $\partial(C_1)$ if n_1 is odd and in $\partial(C_2)$ if n_2 is odd. If n_1 and n_2 are both even or both odd, then S is not in $\partial(C_1) + \partial(C_2)$. These are precisely the cases when $n_1 + n_2$ is even. On the other hand, if just one of n_1 and n_2 is even and the other is odd, then S *is* in $\partial(C_1) + \partial(C_2)$ and $n_1 + n_2$ is odd. Thus S is in $\partial(C_1) + \partial(C_2)$ or not, according to whether $n_1 + n_2$ is odd or not. Turning to $C_1 + C_2$, let n be the number of k-simplexes containing S that are contained in *both* C_1 and C_2. These are the simplexes that cancel out in $C_1 + C_2$. It follows that $n_1 + n_2 - 2n$ is the number of k-simplexes of $C_1 + C_2$ that contain S. Therefore S is in $\partial(C_1 + C_2)$ depending on whether $n_1 + n_2 - 2n$ is odd or even. Since $(n_1 + n_2)$ and $(n_1 + n_2 - 2n)$ are odd or even together, it follows that S belongs to both $\partial(C_1) + \partial(C_2)$ and $\partial(C_1 + C_2)$ or neither. This proves (1).

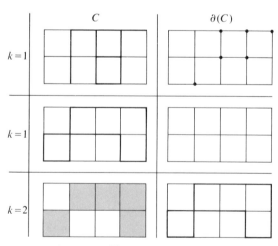

Figure 15.1

Let C be a k-chain. If there is a $(k + 1)$-chain T such that $C = \partial(T)$, then we call C a **k-boundary**. This definition makes sense only if k is 0 or 1. For $k = 2$ we adopt the convention that only $\varnothing$ is a 2-boundary. On the other hand, if $\partial(C) = \varnothing$, then we call C a **k-cycle**. This definition makes sense only if k is 1 or 2. For $k = 0$ we adopt the convention that all 0-chains are 0-cycles. These two special types of chains are crucial in homology.

Exercises

1. Prove that every k-boundary consists of an even number of simplexes. Verify this first for boundaries of simplexes. Complete the proof in two ways:

(a) by induction on the number of simplexes

(b) by using additivity

2. Prove that the sum of two k-boundaries is a k-boundary and the sum of two k-cycles is a k-cycle.

3. For what chains does $\partial(\partial(C))$ exist? Prove that $\partial(\partial(C))$ is always $\varnothing$.

4. Show that every boundary is a cycle.

5. Show that the only 2-cycles are $\varnothing$ and the whole plane.

6. Prove that a 1-chain is a polygonal chain if and only if its boundary consists of exactly two points.

7. Show that topological boundary and chain boundary are the same for 2-chains but not in general for 1-chains.

8. Let the grating $\mathscr{G}$ be enlarged to $\mathscr{G}^+$ by drawing some additional lines. $\mathscr{G}^+$ is called a **subdivision** of $\mathscr{G}$. Show that to every chain C on $\mathscr{G}$ there corresponds a unique chain C^+ on $\mathscr{G}^+$ on the same plane set. Show that $\partial(C^+) = \partial(C)$ to conclude that C^+ is a cycle or a boundary whenever C is a cycle or a boundary.

Let $\mathscr{G}$ be a grating. Two k-chains C_1 and C_2 are called **homologous**, written $C_1 \sim C_2$, if $C_1 + C_2$ is a boundary. Much of the remainder of this book will be devoted to a study of this relationship. We shall see that it possesses a long string of *algebraic* properties that make it easy to deal with and interesting *geometric* properties that make it useful. First, here are the most fundamental algebraic properties of homology.

(a) *reflexive property*: $C \sim C$ for all chains.

(b) *symmetric property*: If $C_1 \sim C_2$, then $C_2 \sim C_1$.

(c) *transitive property*: If $C_1 \sim C_2$ and $C_2 \sim C_3$, then $C_1 \sim C_3$.

(d) *additive property*: If $C_1 \sim C_2$ and $C_3 \sim C_4$, then $C_1 + C_3 \sim C_2 + C_4$.

Here is a proof of the transitive property; the other proofs are left as exercises. Supposing that $C_1 \sim C_2$ and $C_2 \sim C_3$, this means that $C_1 + C_2$ and $C_2 + C_3$ are boundaries of $(k+1)$-chains, say B_1 and B_2, respectively, $\partial(B_1) = C_1 + C_2$, $\partial(B_2) = C_2 + C_3$. Then $B_1 + B_2$ has boundary $\partial(B_1 + B_2) = \partial(B_1) + \partial(B_2) = C_1 + C_2 + C_2 + C_3 = C_1 + C_3$. This proves that $C_1 + C_3$ is a boundary, or that $C_1 \sim C_3$. Note that this argument works only for $k = 0$ or 1. A separate argument is necessary when $k = 2$ (Exercise 10).

These properties of homology should be familiar. They are the standard algebraic properties of equality! For example, (d) is the usual rule: equals added to equals are equal. Only in this case it reads: homologous added to homologous are homologous. It follows from these laws that we obtain three new algebras by replacing equality by homology in the algebras of k-chains already developed. These new algebras are simpler than the original algebras of chains. For example, all boundaries are homologous to the empty chain $\emptyset$. Therefore, when homology replaces equality, all boundaries behave like zero elements.

Exercises

9. Supply proofs for the remaining properties of homology, and verify these additional algebraic properties.

(f) *commutative law*: $C_1 + C_2 \sim C_2 + C_1$ for all k-chains C_1 and C_2

(g) *associative law*: $(C_1 + C_2) + C_3 \sim C_1 + (C_2 + C_3)$ for all k-chains C_1, C_2, and C_3

(h) *zero*: There is a k-chain $\emptyset$, unique up to homology, such that $C + \emptyset \sim C$ for all k-chains C. By unique up to homology we mean that if B is any other k-chain with the zero property ($C + B \sim C$ for all k-chains C), then $B \sim \emptyset$.

(i) *inverses*: Each k-chain C has an inverse D such that $C + D \sim \emptyset$. The inverse is unique up to homology, meaning that if D' is a second inverse for C, then $D' \sim D$.

In other words the k-chains continue to form a group when equality is replaced by homology.

10. Show that for 2-chains homology and equality are the *same*.

11. Let λ_1, λ_2 be 1-chains. Show that if $\lambda_1 \sim \lambda_2$, then λ_1 and λ_2 have the same boundary.

§16 THE FUNDAMENTAL LEMMA

This section is devoted to geometric properties of homology. From Exercise 10 of §15 it follows that homology for 2-chains has no special geometric properties, but for each of the other dimensions homology has an important geometric meaning. Considering one-dimensional homology first, here is the key result.

The Fundamental Lemma

Every 1-cycle is the boundary of exactly two complementary 2-chains.

This is a combinatorial version of the Jordan curve theorem. The 1-cycles are a combinatorial analogue of Jordan curves. They differ from Jordan curves only in that they consist of straight line segments, and they may cross themselves. Topologically a 1-cycle actually may divide the plane into more than two regions, but combinatorially (according to the fundamental lemma) a 1-cycle is the boundary for just two 2-chains (see Figure 16.1). Furthermore these two 2-chains are complementary, just like the regions determined by a Jordan curve. Exactly one of these 2-chains contains the outside face and so is called the outside of the 1-cycle. The other is bounded and is called the inside.

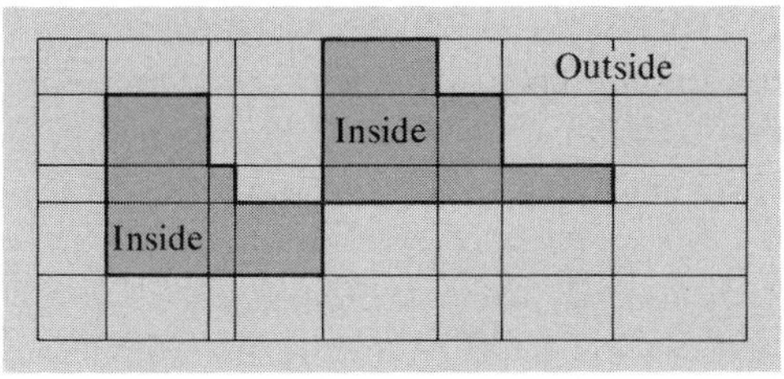

Figure 16.1

The fundamental lemma states an essential property of the plane (and sphere), which, as we shall see, is not possessed by other surfaces. This lemma plays a crucial role in this chapter comparable to the role played by the index lemma in Chapter Two.

Let λ be a 1-cycle. The proof of the fundamental lemma is in two parts. We first prove that there are *at least* two 2-chains with boundary λ, then that there are *at most* two 2-chains with boundary λ. The first part is proved by induction on the number of lines in the grating $\mathscr{G}$. When there are four lines, the grating $\mathscr{G}$ consists of an empty rectangle. Then there are only two 1-cycles, $\varnothing$ and the rectangle itself. The former is the boundary for the 2-chains $\varnothing$ and the whole plane, while the latter is the boundary of the rectangle and the outside face. This disposes of the initial stage of the induction proof.

Assume now that the lemma is true for a given grating $\mathscr{G}$ and let λ be a 1-cycle on a grating $\mathscr{G}^+$ obtained by adding a single new line l to $\mathscr{G}$. We complete the induction proof by proving the lemma for λ. We assume for definiteness that the new line l is horizontal (see Figure 16.2). The idea of the proof is to add something to λ to knock out any edges it has on the line l. The resulting

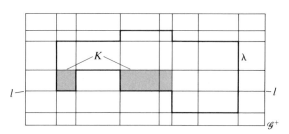

Figure 16.2

cycle will lie, in effect, on $\mathscr{G}$, and thus the induction assumption can be applied. To this end let K be the 2-chain of those faces of $\mathscr{G}^+$ whose lower edges are in λ and rest on l. Then $\partial(K)$ will contain the edges of λ on l that are exactly the ones we would like to eliminate from λ. Let $\mu = \lambda + \partial(K)$; then μ has no edges on l. Furthermore μ is a 1-cycle, since $\partial(\mu) = \partial(\lambda) + \partial(\partial(K)) = \varnothing + \varnothing = \varnothing$. As a 1-cycle on $\mathscr{G}$, by the induction hypothesis, μ is the boundary of at least two different 2-chains S_1 and S_2. Let $T_1 = S_1 + K$ and $T_2 = S_2 + K$. Then $\partial(T_1) = \partial(S_1) + \partial(K) = \mu + \partial(K) = \lambda$ and similarly, $\partial(T_2) = \lambda$. Thus T_1 and T_2 are two 2-chains bounded by λ. These must be distinct, since S_1 and S_2 are distinct.

In the last part of the proof, we show that λ can be the boundary of *at most* two 2-chains. Let T be any 2-chain with boundary λ, and let T_1 be the 2-chain with boundary λ constructed in the first part of the proof. We have $\partial(T + T_1) = \partial(T) + \partial(T_1) = \lambda + \lambda = \varnothing$ so that $T + T_1$ is a 2-cycle. The only 2-cycles are $\varnothing$ and the whole plane (Exercise 5 of §15); therefore either $T + T_1 = \varnothing$ or $T + T_1$ is the whole plane. In the first case $T = T_1$, while in the second case T is complementary to T_1. This completes the proof of the lemma.

Exercises

1. In Figure 16.2 sketch the 1-cycle μ and the two 2-chains S_1 and S_2 that appear in the proof of the fundamental lemma.

2. Prove that two 1-chains are homologous if and only if they have the same boundary.

It follows from the fundamental lemma that homology for two 1-chains means merely that they have the same boundary. If this were all homology meant in general, it would not be worth introducing. However, the fundamental lemma is the starting point for a whole series of related results for homology on different sets.

Definition

Let G be a subset of the plane, and let $\mathcal{G}$ be a grating. Two k-chains on $\mathcal{G}$, C_1 and C_2, are called **homologous in G,** *notated*

$$C_1 \sim C_2 \text{ (in } G)$$

if $C_1 + C_2$ is the boundary of a $(k + 1)$-chain on $\mathcal{G}$ that is contained in the set G.

Homology in a subset G of the plane can be very different from homology in the whole plane. Figure 16.3 contains an example of this. Both parts of the figure display the same pair of 1-chains λ_1 and λ_2. These 1-chains share the same boundary $P + Q$; therefore by the fundamental lemma the cycle $\lambda_1 + \lambda_2$ bounds two complementary 2-chains in the plane. These are shaded in Figure 16.3a. Thus $\lambda_1 \sim \lambda_2$. This changes, however, if we turn our attention from the

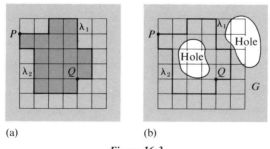

(a)   (b)

Figure 16.3

whole plane to the set G shaded in Figure 16.3b. The cycle $\lambda_1 + \lambda_2$ is not the boundary of a 2-chain (in G) because G doesn't contain either of the 2-chains bounded by $\lambda_1 + \lambda_2$ in the whole plane. What has happened is that G has two holes that prevent $\lambda_1 + \lambda_2$ being a boundary (in G). Therefore λ_1 is not homologous to λ_2 (in G). As this example and the following corollaries to the fundamental lemma show, homology of 1-chains provides an algebraic means for relating 1-chains to "holes" in a given set G.

Corollary 1

Let F be a connected closed set and G the complementary open set. Let $\mathscr{G}$ be a grating. Then every 1-cycle λ in G is the boundary of a chain (in G). In other words, if the open set G has just one hole, then every 1-cycle is still a boundary (in G).

By the fundamental lemma, λ is the boundary of two complementary 2-chains T_1 and T_2. To complete the proof of the corollary we must show that one of these is contained in G. Suppose, on the contrary, that both T_1 and T_2 intersect F. Then the nonempty sets $F \cap T_1$ and $F \cap T_2$ partition F. Every point of F must belong to one set or the other since T_1 and T_2 exhaust the whole plane. Since F is connected, one of these sets contains a point near the other. We conclude that F contains a point of the (topological) boundary between T_1 and T_2. This boundary is λ (Exercise 7 of §15), so this conclusion contradicts the hypothesis that λ is in G. Thus G contains either T_1 or T_2, so that λ is a boundary of a 2-chain (in G).

Corollary 2

Let F be a closed set consisting of two connected components, and let G be the complementary open set. Let $\mathscr{G}$ be a grating. Then (a) $\mathscr{G}$ may contain 1-cycles that are not boundaries (in G); however, (b) any two such 1-cycles in G are homologous (in G).

Figure 16.4 illustrates a typical application of this corollary. The set F is the union of the two components F_1 and F_2 (meaning the set G has two holes). Neither of the 1-cycles λ or μ is the boundary of a 2-chain (in G), because F intersects all the 2-chains for which they are boundaries. However, λ is homologous to μ (in G) since $\lambda + \mu$ is the boundary of the shaded 2-chain (in G).

This example proves part (a) of the corollary. To prove (b), let λ and μ be two 1-cycles in G that are *not* boundaries of 2-chains (in G). By the fundamen-

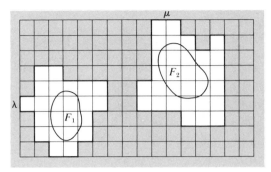

Figure 16.4

tal lemma, λ is the boundary of two 2-chains T_1 and T_2. By the argument of the preceding corollary, the components F_1 and F_2 each lies entirely within either T_1 or T_2. Since λ is *not* a boundary of a 2-chain (in G), it follows that F_1 and F_2 cannot both belong to the same one of the sets T_1 and T_2. Therefore we may as well suppose that $F_1 \subseteq T_1$ and $F_2 \subseteq T_2$. By a parallel argument, μ is the boundary of complementary 2-chains S_1 and S_2, and we may suppose that $F_1 \subseteq S_1$ and $F_2 \subseteq S_2$.

Consider the chain $T_1 + S_1$. Clearly $T_1 + S_1$ is disjoint from F_2, since both T_1 and S_1 are disjoint from F_2. But also $T_1 + S_1$ is disjoint from F_1, since every face of $\mathscr{G}$ that intersects F_1 is contained in both T_1 and S_1 and so cancels out of the sum $T_1 + S_1$. Therefore $T_1 + S_1$ is a 2-chain (in G). Since $\partial(T_1 + S_1) = \lambda + \mu$, this concludes the proof.

Exercises

3. Referring to the proof of Corollary 2, show that $\lambda + \mu$ is the boundary in the whole plane for the *four* 2-chains $S_1 + T_1$, $S_1 + T_2$, $S_2 + T_1$, and $S_2 + T_2$. Does this contradict the fundamental lemma?

4. Consider a compact set F that is the union of three components F_1, F_2, and F_3. Let $\mathscr{G}$ be a grating containing three 1-cycles λ_1, λ_2, and λ_3 such that F_1 is inside λ_1, F_2 is inside λ_2, and F_3 is inside λ_3, but other components are outside these cycles. Draw a picture to illustrate this situation.

The following exercises refer to your drawing for Exercise 4.

5. Show that none of the cycles λ_1, λ_2, or λ_3 is a boundary in G, where G is the open set complement of F. Prove that no two of these cycles are homologous in G.

6. Show that $\lambda_1 + \lambda_2 \sim \lambda_3$ (in G).

7. Show that any 1-cycle on $\mathcal{G}$ is homologous in G to exactly one of the cycles $\emptyset$, λ_1, λ_2, or $\lambda_1 + \lambda_2$.

A set of 1-cycles on $\mathcal{G}$ is called a **representing set for G** if every 1-cycle on $\mathcal{G}$ is homologous (*in G*) to exactly one of the cycles of the set. Exercise 7 asserts that $\{\emptyset, \lambda_1, \lambda_2, \lambda_3\}$ is a representing set for the G of that exercise. A set of 1-cycle is a **homology basis for G** when every 1-cycle on $\mathcal{G}$ is homologous (*in G*) to exactly one of the following cycles: (a) $\emptyset$, (b) one of the cycles in the basis, or (c) a sum of cycles in the basis. Exercise 7 also asserts that $\{\lambda_1, \lambda_2\}$ constitutes a homology basis for G.

---

8. Consider a set F that has four components, and find a representing set and a homology basis for the complementary set G on a suitable grating.

9. Suppose F has n components. Make a conjecture about the numbers of cycles in a representing set and a homology basis.

Turning to the geometric interpretation of zero-dimensional homology, we find that homology for 0-chains is an algebraic reflection of the connectedness of sets. More precisely we have the following theorem, whose proof is an application of the polygonal chain theorem.

Theorem

Let G be an open set and let P and Q be two points of G. Then $P \sim Q$ (in G) if and only if P and Q can be connected by a polygonal chain in G.

Exercises

10. Prove the theorem just stated above.

§17 ALEXANDER'S LEMMA

This section contains the last two lemmas needed in the proof of the Jordan curve theorem. First consider the following connection problem. Let F_1 and F_2 be compact sets and G_1 and G_2 their complements. Suppose that the points P and Q may be connected in G_1 *and* G_2. When may they be connected in $G_1 \cap G_2$? Alexander's lemma gives a sufficient condition.

Alexander's Lemma

Let G_1 and G_2 be open sets with compact complementary sets (holes) F_1 and F_2, respectively. Let P and Q be points such that

$$P \sim Q \ (in \ G_1) \quad and \quad P \sim Q \ (in \ G_2)$$

Let λ_1, and λ_2 be chains in G_1 and G_2, respectively, such that $\partial(\lambda_1) = \partial(\lambda_2) = P + Q$. Let $\mathcal{G}$ be a grating containing both λ_1 and λ_2. If

$$\lambda_1 \sim \lambda_2 \ (in \ G_1 \cup G_2)$$

then on a suitable subdivision $\mathcal{G}^+$ of $\mathcal{G}$,

$$P \sim Q \ (in \ G_1 \cap G_2)$$

The two parts of Figure 17.1 illustrate the problem described above. In each part the points P and Q are connected by polygonal paths λ_1 and λ_2 that avoid the sets F_1 and F_2, respectively. (In other words, λ_1 is in G_1 and λ_2 is in G_2.) In Figure 17.1a P and Q *can* be connected by a path avoiding both F_1 and F_2 (in other words, by a path in $G_1 \cap G_2$), but in Figure 17.1b this is

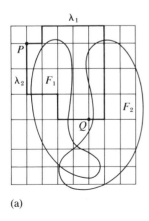

(a)

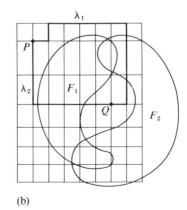

(b)

Figure 17.1

impossible. In Figure 17.1a $\lambda_1 + \lambda_2$ is a boundary in the plane minus the set $F_1 \cap F_2$ (in other words, in $G_1 \cup G_2$). It is this circumstance, lacking in Figure 17.1b, that enables us, in the following proof, to find a path connecting P and Q in $G_1 \cap G_2$.

The idea of the proof is to add something to λ_1 to help it avoid F_2. The problem is to do this without pushing λ_1 into F_1. The solution of this difficulty is provided by the following lemma.

Lemma

Let E_1 and E_2 be disjoint compact sets. Then there is a constant $\varepsilon > 0$ such that if R is a point of E_1 and S is a point of E_2, then

$$\|R - S\| > \varepsilon$$

The proof of this lemma is a familiar compactness argument. Supposing on the contrary that for every $\varepsilon > 0$ there is a pair of points $R \in E_1$, and $S \in E_2$ such that $\|R - S\| < \varepsilon$, then in particular we can find sequences of points $\{R_n\} \subseteq E_1$ and $\{S_n\} \subseteq E_2$ such that $\|R_n - S_n\| < 1/n$. It follows by compactness that one of the sets E_1 or E_2 contains a point near the other. Since E_1 and E_2 are closed, this point is in both sets, contradicting the assumption that E_1 and E_2 are disjoint, and proving the lemma.

Returning to the proof of Alexander's lemma, by assumption $\lambda_1 + \lambda_2$ is the boundary of a 2-chain T in $G_1 \cup G_2$. Then the compact sets $T \cap F_1$ and $T \cap F_2$ are disjoint. By the preceding lemma there is a constant $\varepsilon > 0$ so that points of these two sets are always separated by a distance at least ε. Therefore by subdivision of the grating $\mathcal{G}$ we can create a grating $\mathcal{G}^+$ such that no face of the grating intersects both $T \cap F_1$ and $T \cap F_2$. All that is required is the addition of sufficient lines so that the sides of the faces of the new grating are all less than $\varepsilon/2$. (See Figure 17.2.) Supposing this done, we now replace λ_1, λ_2, and T by their subdivisions on $\mathcal{G}^+$.

Let S be the 2-chain on $\mathcal{G}^+$ of those faces of T that intersect F_2. By construction S is disjoint from F_1. Let $\lambda = \lambda_1 + \partial(S)$. Since $\partial(\lambda) = \partial(\lambda_1) + \partial(\partial(S)) = P + Q$, λ connects P and Q. By proving that λ is disjoint from F_1 and disjoint from F_2, we will conclude the proof of the lemma. Observe first that both λ_1 and S are disjoint from F_1; therefore λ is disjoint from F_1. Next, note that $T + S$ is disjoint from F_2 since it is precisely those faces of T that intersect F_2 that are in S and so are cancelled in this sum. Now

$$\partial(S + T) = \partial(S) + \partial(T) = \partial(S) + \lambda_1 + \lambda_2 = \lambda + \lambda_2$$

Therefore $\lambda = \partial(S + T) + \lambda_2$. We conclude that λ is disjoint from F_2, because both $S + T$ and λ_2 are disjoint from F_2.

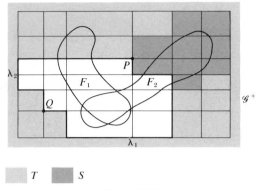

Figure 17.2

As an application of Alexander's lemma, we can prove the following theorem, which is needed in any event for the proof of the Jordan curve theorem.

The Jordan Curve Theorem for Paths

Paths do not divide the plane.

Let γ be a path. Let P and Q be two points not on γ. Suppose, contrary to what we wish to prove, that P and Q cannot be connected in the open set G, the complement of γ. We deduce a contradiction by the method of bisection. Let T be a topological transformation from the unit interval $[0, 1]$ onto γ. Let γ_1 and γ_2 be the two "halves" of γ that correspond by T to the two halves $[0, \frac{1}{2}]$ and $[\frac{1}{2}, 1]$ of the unit interval. We shall prove that one of these two halves, γ_1 or γ_2, also separates P from Q. Granted this for the moment, we take this half and halve it in turn by considering quarters of the unit interval. By the argument we are assuming, one of these "quarters" of γ also divides P from Q. The bisection continues and an "eighth" of γ is found that divides P from Q. The sequence of subintervals of the unit interval chosen in this manner converges to a point (Exercise 1). By continuity, the corresponding segments of γ converge to a point S on γ. Thus the points P and Q are divided by portions of γ contained in any neighborhood of S. This is absurd, so the original hypothesis that γ divides P and Q must be false.

Returning to the original bisection of γ, we complete the proof by showing that one of the halves alone divides P from Q. Suppose, on the contrary, that neither γ_1 nor γ_2 divides P from Q. (This is a proof by contradiction within a proof by contradiction!) Then there is a polygonal path λ_1 connecting P and Q without intersecting γ_1 and a corresponding path λ_2 that avoids γ_2. We may assume that λ_1 and λ_2 are 1-chains on a suitable grating surrounding the points P and Q and the *closed* sets γ_1 and γ_2. Let G_1 and G_2 be the complements of γ_1 and γ_2, respectively. Then λ_1 is in G_1, λ_2 is in G_2, and $\partial(\lambda_1) = \partial(\lambda_2) = P + Q$. Since $G_1 \cup G_2$ is the whole plane minus the single point that is the intersection of γ_1 and γ_2, it follows by the first corollary of the second fundamental lemma that $\lambda_1 \sim \lambda_2$ in $G_1 \cup G_2$. Thus by Alexander's lemma $P \sim Q$ in $G_1 \cap G_2$. But this contradicts our assumption that P and Q are divided by γ, so we conclude that one of the paths γ_1 or γ_2 does divide P from Q.

Exercises

1. Let $[a_n, b_n]$ be a sequence of intervals each interval half as long as its predecessor. Use a compactness argument to prove that there is a point x such that $\lim a_n = \lim b_n = x$.

2. Use the techniques of this section to prove the following theorem: a cell does not divide the plane.

---

§18 PROOF OF THE JORDAN CURVE THEOREM

Let $\mathscr{J}$ be a Jordan curve. The proof of the theorem is divided into four parts.

(I) $\mathscr{J}$ DIVIDES THE PLANE INTO AT MOST TWO PIECES

Let P, Q, and R be three points in $\mathscr{J}$. Suppose that P and Q are divided by $\mathscr{J}$ and that Q and R are divided by $\mathscr{J}$, as in Figure 18.1. We shall prove that P and R are not divided by $\mathscr{J}$, that they can be connected in $\mathscr{J}'$. Let A and B be any two points on $\mathscr{J}$. They determine two paths, γ_1 and γ_2, whose union is $\mathscr{J}$ and whose intersection is the pair of points A and B. By the Jordan curve theorem for paths, neither γ_1 nor γ_2 alone divides the plane. Therefore there is a grating $\mathscr{G}$ and 1-chains $\lambda_1, \lambda_2, \mu_1,$ and μ_2 such that $\partial(\lambda_1) = \partial(\lambda_2) = P + Q$, $\partial(\mu_1) = \partial(\mu_2) = Q + R$, μ_1 and λ_1 do not intersect γ_1, and μ_2 and λ_2 do not intersect γ_2. Let G_1 and G_2 be the open sets complementary to γ_1 and γ_2.

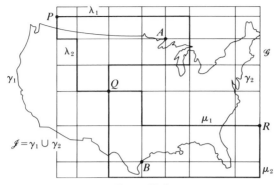

Figure 18.1

The 1-cycle $\lambda_1 + \lambda_2$ cannot be a boundary in $G_1 \cup G_2$ or else by Alexander's lemma P and Q could be connected in $G_1 \cap G_2$ contrary to our hypothesis. Similarly, $\mu_1 + \mu_2$ cannot be a boundary in $G_1 \cup G_2$. Since $G_1 \cup G_2$ is the complement of the two-point set $\{A, B\}$, it follows by the second corollary to the fundamental lemma that

$$(\lambda_1 + \lambda_2) \sim (\mu_1 + \mu_2) \quad (\text{in } G_1 \cup G_2)$$

In other words, $\lambda_1 + \lambda_2 + \mu_1 + \mu_2$ is a boundary in $G_1 \cup G_2$. We now apply Alexander's lemma to the two chains $(\lambda_1 + \mu_1)$ and $(\lambda_2 + \mu_2)$. Each of these connects P with R; the first is in G_1 while the second is in G_2. Furthermore, $(\lambda_1 + \mu_1) + (\lambda_2 + \mu_2)$ is a boundary in $G_1 \cup G_2$. Therefore we conclude that P and R can be connected in $G_1 \cap G_2 = \mathscr{J}'$. Thus $\mathscr{J}$ does not divide P from R.

(II) $\mathscr{J}$ DIVIDES THE PLANE INTO AT LEAST TWO PIECES

We must now produce two points that are divided by $\mathscr{J}$. To begin, let A and B be any two points on $\mathscr{J}$. As in (I), they divide $\mathscr{J}$ into two paths γ_1 and γ_2. Choose a square σ around A small enough so that B is outside σ. The paths γ_1 and γ_2, since they connect a point inside σ with a point outside σ, must intersect σ, according to the Jordan curve theorem for squares (Exercise 9 of §13). Since σ and $\mathscr{J}$ must cross, we expect that σ will contain points on "both sides" of $\mathscr{J}$, points that cannot be connected without crossing $\mathscr{J}$, and this is what we shall prove.

To accomplish this, choose a grating $\mathscr{G}$ on which σ is a 1-chain and such that no edge of σ intersects both sets γ_1 and γ_2 (the lemma and argument used

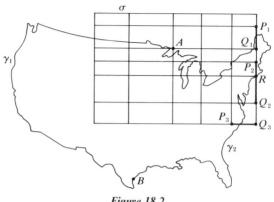

Figure 18.2

in the proof of Alexander's lemma are relevant here). Let λ be the 1-chain consisting of those edges of σ that intersect γ_2 (see Figure 18.2). By construction λ is not empty. At the same time λ cannot be all of σ, since σ also intersects γ_1 (but with a different set of edges). Thus $\partial(\lambda)$ cannot be empty but consists of an even number of vertexes (Exercise 1 of §15). Let these vertexes be arranged in pairs $P_1, Q_1 ; P_2, Q_2 ; \ldots ; P_n, Q_n$. None of these vertexes is on $\mathscr{J}$ since any vertex of σ on $\mathscr{J}$ will be in *two* edges of λ (see the point R in Figure 18.2). We claim that at least one of these pairs is separated by $\mathscr{J}$. Intuitively this is clear, because γ_2 crosses σ an *odd* number of times. The proof is by contradiction. Suppose each of these pairs of points is connected by a path that does not intersect $\mathscr{J}$. Then on a suitable subdivision $\mathscr{G}^+$ these polygonal paths are all 1-chains, say $\mu_1, \mu_2, \ldots, \mu_n$, where $\partial(\mu_1) = P_1 + Q_1$, $\partial(\mu_2) = P_2 + Q_2, \ldots,$ $\partial(\mu_n) = P_n + Q_n$. Let $\mu = \mu_1 + \mu_2 + \mu_3 + \cdots + \mu_n$; then $\partial(\mu) = \partial(\lambda)$. Thus $\lambda + \mu$ is a 1-cycle. Let G_1 and G_2 be the open sets complementary to γ_1 and γ_2. Since both μ and λ do not intersect γ_1, $\mu + \lambda$ is a 1-cycle in G_1. By the first corollary to the second fundamental lemma, $\lambda + \mu$ is a boundary in G_1.

The 1-chain $\lambda + \mu + \sigma$ is also a 1-cycle. By construction, $\lambda + \sigma$ does not intersect γ_2, since the edges of σ that do intersect γ_2 are cancelled in this sum. Therefore $\lambda + \mu + \sigma$ is a 1-cycle in G_2. Again by the first corollary to the fundamental lemma, $\lambda + \mu + \sigma$ is a boundary in G_2. It follows that

$$\sigma = (\lambda + \mu) + (\lambda + \mu + \sigma)$$

is a boundary in $G_1 \cup G_2$. But $G_1 \cup G_2$ is the whole plane minus the two points A and B. The rectangle σ was chosen just so σ would *not* be a boundary in $G_1 \cup G_2$. This is the desired contradiction.

(III) $\mathscr{J}$ FORMS THE BOUNDARY OF BOTH PIECES

We now know that $\mathscr{J}$ divides the plane into two components R_1 and R_2. Since $\mathscr{J}$ is closed, R_1 and R_2 are open and hence do not contain their boundaries: $b(R_1) \subseteq R_2 \cup \mathscr{J}$ and $b(R_2) \subseteq R_1 \cup \mathscr{J}$. But points of R_1 cannot be near R_2 (and vice versa) since these are open sets; therefore both $b(R_1)$ and $b(R_2)$ are subsets of $\mathscr{J}$. In (II) we showed that an arbitrary point A of $\mathscr{J}$ is near both R_1 and R_2 (since the rectangle σ, *which could be arbitrarily small*, contained points of R_1 and R_2). Thus $\mathscr{J}$ is the complete boundary of R_1 and R_2.

(IV) ONE OF WHICH IS BOUNDED, THE OTHER UNBOUNDED

Since $\mathscr{J}$ is bounded, it can be contained in a disk. Exactly one of the regions R_1 or R_2 will contain the connected set of points outside this disk and therefore be unbounded. The other region determined by $\mathscr{J}$ will be bounded. This completes the proof of the Jordan curve theorem.

The proof of this theorem was long but has served to introduce the concept of homology. You certainly appreciate now the neat way homology handles the various delicate questions of bounding and connection involved in the Jordan curve theorem. However, homology will prove itself useful in a far larger range of problems. When we return to homology in Chapter Five and apply it to arbitrary surfaces, we will find it a tool equal in power to all the combinatorial devices we have studied thus far—Euler's formula, the Sperner and index lemmas, *and* plane homology—combined!

Exercise

Let $\mathscr{D}$ be a closed disk, and let f be a topological transformation from $\mathscr{D}$ to a set R. The boundary of $\mathscr{D}$, $b(\mathscr{D})$, is a circle; therefore $f(b(\mathscr{D}))$ is a Jordan curve. Prove that $f(I(\mathscr{D}))$ is the inside of $f(b(\mathscr{D}))$ in order to conclude that every cell consists of a Jordan curve as boundary together with its inside.

Notes. The discussion of the Jordan curve theorem given here is based on Newman [25]. More advanced treatments, including generalizations to higher dimensions, are given in the books by Eilenberg and Steenrod [7], Greenberg [9], and Vick [29].

four

Surfaces

§19 EXAMPLES OF SURFACES

You have now been introduced to several techniques of combinatorial topology—the Euler characteristic, winding numbers, index theorems, and plane homology—as applied to some problems in the plane: fixed points of continuous transformations, singular points of vector fields, and Jordan curves. The object of the remainder of the book is to expand these techniques until they apply to the same aspects of other topological spaces. This chapter presents the topological spaces that we hope to study in this way: the surfaces. Surfaces were the first topological spaces to be studied in detail. As a result they are among the best understood of topological spaces. This circumstance together with the many applications of surfaces to other branches of mathematics has made their theory a model for the investigation of other topological spaces. Unfortunately, no other class of topological spaces has turned out to be as well behaved.

Surfaces are those topological spaces in which every point has a neighborhood that is topologically equivalent to an open disk. To put it dramatically, people living at a point on a surface, if their perception is limited to a small neighborhood of their point, are unable to distinguish their situation from that of people actually living on a plane. Our situation, living on the earth, fits this description exactly (and indeed the sphere is a surface). The simplest example of a surface is naturally the plane itself or, topologically equivalent, an open disk. More generally, any open subset of the plane is a surface.

Other examples may be constructed combinatorially by gluing disks together according to the program stated in §1. The simplest surface of this type is the **cylinder** obtained from a single disk in the form of a rectangle by gluing together a pair of opposing edges. Figure 19.1a shows a rectangle before gluing. In Figure 19.1b the sides labeled *a* have been glued together to form a cylinder. With the exception of the top and bottom edges, every point on the cylinder has a neighborhood equivalent to a disk, for example the point *P* in Figure 19.1. Due to the presence of the exceptional points, the cylinder is called a **surface with boundary** (see §22).

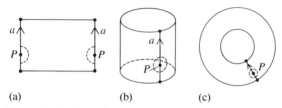

Figure 19.1   The cylinder. (a) Rectangular model.
(b) Space model. (c) Cylinder flattened into an annulus.

As a fundamental principle of our study, we will seek a model in the plane for every surface. The cylinder is unusual in this respect since it can actually be topologically transformed (flattened) into a plane set (Figure 19.1c). However, a still more useful model can be obtained from the original rectangle. This is accomplished by defining a special topology on the rectangle, a topology that pretends that the rectangle has already been glued together. The neighborhoods of this topology for most points are the usual circular disks inside the rectangle, but for points on the edges *a* the neighborhoods consist of two half disks around symmetrical points across the rectangle (Figure 19.1a). A more formal definition will be given in §20. Informally it is clear that the rectangle is given its usual topology by this device, except that symmetrical points on the edges *a* have the same neighborhoods: the neighborhoods they would have if they were glued together. The symmetrical pairs of points are said to be **topologically identified**. Topological identification has the same effect as gluing or pasting but is a great deal more convenient, because it permits us to contemplate the resulting topological space in the plane rather than in space.

Our next example of a surface is the sphere (see Figure 19.2). In order to find a plane model of the sphere we must find a means for making a sphere from a plane set by sewing or pasting. This turns out to be even simpler than the cylinder, for the sphere can be regarded as the result of taking a single cell in the form of a 2-gon (Figure 19.2a) and sewing the cell to itself along its two sides (Figure 19.2b). If we were actually to perform this gluing operation, the result would be like a football or a purse, figures that, of course, are topologically equivalent to a sphere.

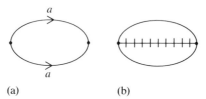

(a)   (b)

Figure 19.2   The sphere. (a) Plane model.
(b) Space model.

The next example that naturally presents itself is the **torus**: the surface that results when both pairs of opposite sides of a rectangle are identified (Figure 19.3a). If the sewing indicated by these identifications is carried out, one obtains a hollow doughnut-shaped surface (Figure 19.3c). Note that all four vertexes of the original rectangle end up identified to a single point. On this account the neighborhoods of this point in the plane model consist of quarter disks about each of the corner points of the rectangle.

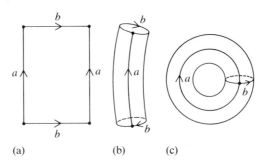

(a)   (b)   (c)

Figure 19.3   The torus. (a) Plane model.
(c) Space model.

One of the advantages of the plane model of the torus (or any other surface) is that it enables us to see the whole surface at once, while a perspective drawing of a space model inevitably hides some portions of the surface. For example, Figure 19.4 shows the same closed path drawn on both models of the torus. On the space model parts of the path are hidden, while the whole path is visible on the plane model. Such paths arise in the solution of differential equations on the torus. Note that when the path goes off an edge of the plane model, it immediately reenters the model at the symmetrically placed point on the opposite edge.

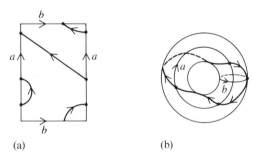

(a)   (b)

Figure 19.4   A closed path on the torus.
(a) Plane model. (b) Space model.

To continue this sequence of examples, we next introduce the **two-holed torus**. Rather than do this directly, we shall consider a general technique for constructing new examples of surfaces from old. Given two surfaces, the idea is to cut out a disk from each and sew the two resulting surfaces together along the boundary of the cutouts. The surface that results is called the **connected sum** of the two original surfaces. Figure 19.5 shows this process applied to two tori. The result is a two-holed torus. If we now take the connected sum of a torus and a two-holed torus, we obtain a three-holed torus. Further connected sums result in four-holed tori, five-holed tori, and so on. You should immediately ask, What are the plane models of these surfaces? The plane models are found by performing the operation of connected sum with the plane models of the surfaces, the result being a plane model of the connected sum. Figure 19.6 shows this process applied to two tori. Note that the disks are cleverly cut from corners of the tori. The holes (h = hole) can then be stretched open so that the sewing can be performed in the plane (Figure 19.6a and b). The result is an octagonal model of the two-holed torus (Figure 19.6c). Figure 19.7 shows how this model actually yields a two-holed torus when the edges of the octagon are sewn together as indicated. Note that all the vertexes of the octagon are sewn together to a single point on the two-

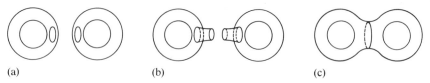

Figure 19.5 The connected sum of two tori worked out in space. (a) Two tori with disks cut out. (b) Stretching the holes before sewing. (c) The connected sum.

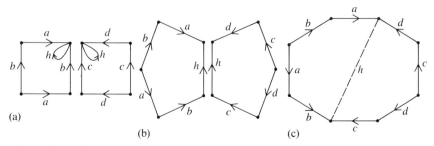

Figure 19.6 The connected sum of two tori worked out in the plane. (a) Two tori with disks cut out. (b) Stretching the holes before sewing. (c) The connected sum.

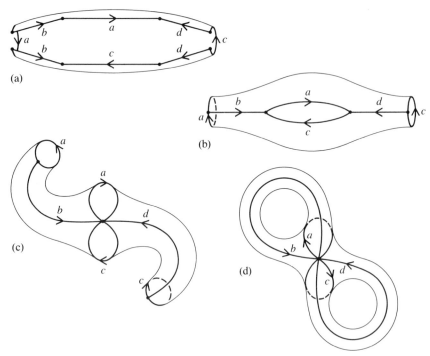

Figure 19.7 Sewing a two-holed torus together from the plane model.

holed torus. In a similar fashion we obtain models for the three-holed torus, four-holed torus, and so forth. In general, the n-holed torus will be a $4n$-gon whose edges are identified in pairs, these pairs themselves arranged in inter-twined pairs of pairs of edges around the polygon (Figure 19.8).

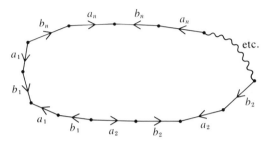

Figure 19.8   An n-holed torus.

Exercises

1. What familiar surfaces are represented by the following plane models?

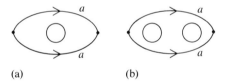

(a)   (b)

2. Show that the operation of connected sum is commutative and associa-tive. Show that the sphere is an identity for connected sums.

3. Find plane models for the connected sum of a cylinder and a torus, and for the connected sum of two cylinders.

4. For each of the following surfaces, sketch the space model together with the indicated path.

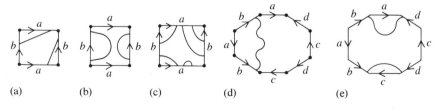

(a)   (b)   (c)   (d)   (e)

Paralleling the family of surfaces already introduced is another world composed of the twisted surfaces. The most famous of these is the **Möbius strip**, or **twisted cylinder**, invented by Möbius in 1858. Like the cylinder, the Möbius strip is formed by identifying opposite sides of a rectangle. The difference is that now the rectangle is twisted before identification so that opposite vertexes are sewn together (Figure 19.9a). This is indicated in the plane model (Figure 19.9b) by drawing arrows on the identified edges pointing in opposite directions. The figure also shows a point with a neighborhood on the identified edge. The Möbius strip, like the cylinder, is not a true surface but a **surface with boundary**. Unlike the cylinder, the Möbius strip has just one boundary curve. The two unidentified sides of the rectangle are actually joined by the identification of opposing vertexes into a single boundary edge. You may want to construct a space model of a Möbius strip to be convinced of this property. This is your last chance, because succeeding examples of surfaces cannot be constructed at all in three-dimensional space.

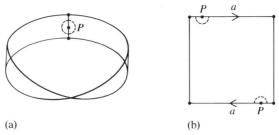

(a)   (b)

Figure 19.9   The Möbius strip. (a) Space model.
(b) Plane model.

The property of the Möbius strip of greatest interest to its inventor was its one-sidedness. By this is meant that it is possible for an insect (say) to walk over the entire surface without crossing an edge, a trip that is not possible on the ordinary cylinder. Another extraordinary property is the behavior of the Möbius strip when cut down the middle, parallel to the edge. The cylinder so sliced divides into two cylinders (Figure 19.10a). The same cut on a Möbius strip *appears* to divide the plane model into two pieces, but these two pieces are actually connected by parts of their boundaries, which remain to be identified. If we sew those pieces back together (the second piece must be turned over in order to do this), then we see that the Möbius strip has become a single cylinder (Figure 19.10b). It was Möbius, incidentally, who invented these plane models.

By analogy with the sphere we also have the **twisted sphere**, or **projective plane** (Figure 19.11). This is a 2-gon with sides twisted and identified. Like the

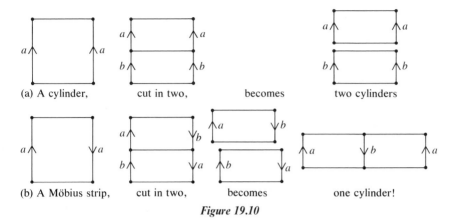

(a) A cylinder,   cut in two,   becomes   two cylinders

(b) A Möbius strip,   cut in two,   becomes   one cylinder!

Figure 19.10

ordinary sphere, this is a surface without boundary. The projective plane can be thought of as a circular disk with diametrically opposite points identified. It is impossible to carry out this identification in three-dimensional space.

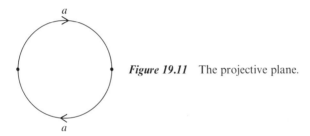

Figure 19.11   The projective plane.

Another surface that cannot be constructed in space is the **twisted torus**, or **Klein bottle** (Figure 19.12a), a figure almost as celebrated as the Möbius strip. It is possible to perform one of the identifications stipulated in the plane model (Figure 19.12b), but it is not possible to make the second identification

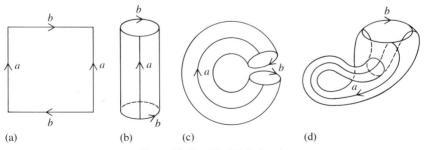

(a)   (b)   (c)   (d)

Figure 19.12   The Klein bottle.

directly, because the orientations of the boundary curves don't match (Figure 19.12c). It is necessary to bring one end of the tube through the side of the surface. The resulting figure in space (Figure 19.12d) is not really a Klein bottle because of the unavoidable self-intersection. The figure may still help in visualizing the surface. Ignoring this self-intersection, the Klein bottle is a very peculiar bottle. There is no boundary or mouth to the bottle, so it must be considered a closed bottle, but it does not divide space into an inside and an outside.

Using connected sums we can produce an infinite series of twisted surfaces to parallel the infinite series of tori. For this purpose we use the projective plane rather than the Klein bottles, since the Klein bottle is itself the connected sum of two projective planes! This is demonstrated in Figure 19.13, where we have first worked out the connected sum of two projective planes and then shown how this connected sum equals the Klein bottle.

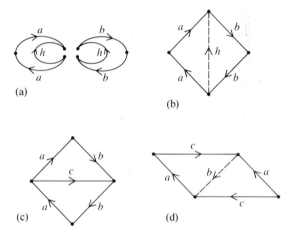

Figure 19.13 Formation of a Klein bottle. (a) Two projective planes with holes cut out. (b) The projective planes sewn together to form a connected sum. (c) The connected sum cut into two triangles. (d) The two triangles sewn together again (one triangle turned over) to form a Klein bottle.

More generally, Figure 19.14 shows the plane model for the connected sum of n projective planes.

Figure 19.14   The connected sum of n projective planes.

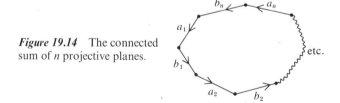

Exercises

5. Investigate what happens when a Möbius strip is cut parallel to its boundary at a distance one third the width of the strip from the boundary. Carry out your investigation with paper and scissors and with the plane model.

6. Prove that a projective plane with a hole cut out is a Möbius strip by cutting up the model of the projective plane with a hole and sewing it back together again.

7. Prove that two Möbius strips sewn together along their boundaries make a Klein bottle. This can be done two ways:

(a) using the usual model of the Möbius strip (Figure 19.9)

(b) using the model suggested by Exercise 6

8. Determine all the different ways that the sides of a rectangle can be identified in pairs. In each case decide which of the surfaces introduced in this section the rectangle represents. Repeat this exercise with a triangle.

Among the many questions that can be asked about a given surface, some of the most important grow out of the problems in the plane to which the first three chapters of this book were devoted. Let us illustrate this with the example of the torus.

It is possible to divide the torus into faces, edges, and vertexes just as we divided the sphere in Chapter One. One such **toroidal polyhedron** is given in Figure 19.15. On the sphere the number of faces minus the number of edges plus the number of vertexes is always two. This is expressed by saying that the Euler characteristic of the sphere is two. The question is whether this sum is always the same on the torus as well. If so, then Figure 19.15 shows that the Euler characteristic of the torus is zero. (The existence of the Euler characteristic for all surfaces is proved in Chapter Five.)

In another direction, consider a vector field on the torus. The important vector fields are the **tangent vector fields,** in which each vector $V(P)$ lies in the tangent plane to the torus at P. An example is given in Figure 19.16a. Just as in the plane, a tangent vector field is associated with a system of differential equations on the torus whose solutions form a family of paths called the **phase portrait of the vector field** (Figure 19.16b). The plane model is particularly useful in visualizing these phase portraits (Figure 19.16c). Here the most obvious feature is the saddle point at P. Less obvious are the two centers at Q

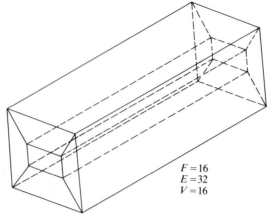

$$F = 16$$
$$E = 32$$
$$V = 16$$

Figure 19.15

and R. Finally, although it takes an effort to see it, there is another saddle point at the corners S. No mere mathematical curiosity, differential equations on the torus arise from doubly periodic differential equations in the plane. (Tangent vector fields on surfaces are discussed in Chapter Five.)

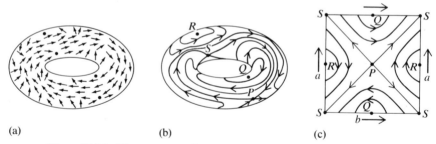

(a)   (b)   (c)

Figure 19.16   Three aspects of a vector field. (a) A tangent vector field. (b) Phase portrait in space. (c) Phase portrait on the plane model.

Continuous transformations of surfaces are naturally important in topology. Their theory for arbitrary surfaces is not related to vector fields as was the case in the plane. Continuous transformations need not have any fixed points, as the example of a rotation of the torus shows. Nevertheless, a study of fixed points can be very useful. (Continuous transformations of surfaces are studied in Chapter Six.)

In this way the first two chapters serve as an outline for the second half of this book. In the remainder of this chapter we present the formal theory of surfaces, which, together with the homology of Chapter Three, provides the foundation for this plan.

Exercises

9. Here are a few more toroidal polyhedra. For each, check that $F - E + V = 0$. Draw the plane model corresponding to Fig. 19.15. Could any of these be called regular polyhedra?

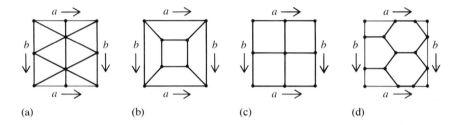

(a)   (b)   (c)   (d)

10. Given two surfaces with Euler characteristics A and B, show that the Euler characteristic of the connected sum of the two surfaces is $A + B - 2$.

11. For each of the following phase portraits, find and classify all the critical points. In each case and in the case of the phase portrait of Figure 19.16, compute the sum of the indexes of the critical points.

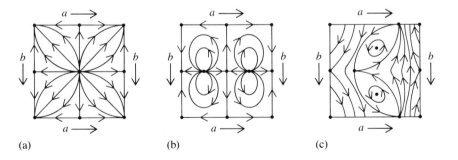

(a)   (b)   (c)

12. The Jordan curve theorem also suggests an interesting question for any surface: does there exist a Jordan curve on the given surface that does *not* divide the surface into two pieces? We will be unable to treat this question

completely; as a start, however, consider the following Jordan curves on various surfaces. In each case tell whether the curve does or does not divide the surface.

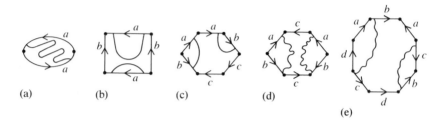

(a)   (b)   (c)   (d)

(e)

13. Show that it is possible to cut the torus along two Jordan curves simultaneously and still not divide the surface. Show that the same is possible on the n-holed torus with $2n$ cuts. What is the situation with the projective plane and the other twisted surfaces?

§20 THE COMBINATORIAL DEFINITION OF A SURFACE

We begin by defining carefully the crucial process of identification.

Definition

Let $\mathscr{P}$ be a collection of polygons, and let $a_1, a_2, \ldots, a_n$ be a set of edges from these polygons. These edges are termed **identified** when a new topology is defined on $\mathscr{P}$ as follows: (a) each edge is assigned a direction from one endpoint to the other and placed in topological correspondence with the unit interval in such a way that the initial points of all the edges correspond to 0 and the final points correspond to 1; (b) the points on the edges $a_1, a_2, \ldots, a_n$ that all correspond to the same value from the unit interval are treated as a single point; and (c) the neighborhoods of the new topology on $\mathscr{P}$ are the disks entirely contained in a single polygon plus the unions of half disks whose diameters are matching intervals around corresponding points on the edges $a_1, a_2, \ldots, a_n$.

Figure 20.1 gives an example in which $\mathscr{P}$ consists of four polygons and a total of five edges are identified. The direction of these edges is indicated by an arrow. The "corresponding points" referred to in the definition are the

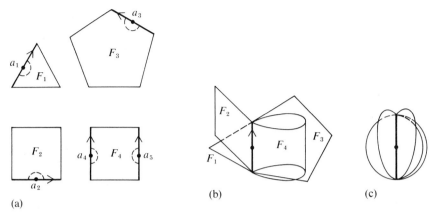

Figure 20.1 Identification at edges. (a) Four polygons identified at five edges.
(b) The identification carried out in space. (c) A typical neighborhood (blown up).

sets of points lying a given proportion of the way along the edges, for example, all the points lying one half of the way along the edges, as in Figure 20.1. This definition applies to all the examples discussed in §19. Note that two or more edges can lie on the same polygon. The next definition concerns identification at vertexes.

Definition

Let $\mathscr{P}$ be a set of polygons, and let $P_1, P_2, \ldots, P_n$ be a collection of vertexes from these polygons. These vertexes are said to be **identified** *when a new topology is defined on $\mathscr{P}$ in which this collection of vertexes is treated as a single point and the neighborhoods are defined to be disks completely contained in a single polygon plus the unions of portions of disks around each of the points $P_1, P_2, \ldots, P_n$. In case any of the edges meeting at one of these vertexes is also identified, the sectors forming a neighborhood at the "point" $\{P_1, P_2, \ldots, P_n\}$ must contain matching intervals from these edges.*

Figure 20.2 gives an example in which $\mathscr{P}$ consists of six polygons and six vertexes are identified. In this example, some edges are identified also so that both processes of identification can be viewed together.

With these preliminaries established, we can proceed to a formal definition of surfaces. From a combinatorial viewpoint, all topological spaces should be constructed by simple operations from simple pieces. In the case of surfaces the pieces are *polygons* and the operations are *topological identifica-*

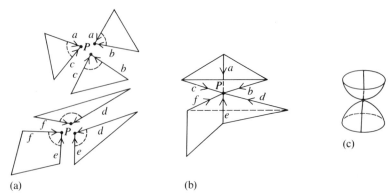

Figure 20.2   Identification at vertexes. (a) Six polygons identified at a vertex. (b) The identification carried out in space. (c) A neighborhood of the vertex.

tions. However, topological identification can produce points whose neighborhoods are not disks, as Figure 20.1 and 20.2 demonstrate. The following definition is designed to exclude such pathological neighborhoods and so preserve in combinatorial form the intuitive notion that a surface is a topological space for which every point has a neighborhood equivalent to a disk.

Definition

*A topological space is **triangulable** if it can be obtained from a set of triangles by the identification of edges and vertexes subject to the restriction that any two triangles are identified either along a single edge or at a single vertex, or are completely disjoint. A **surface** is a triangulable space for which in addition (a) each edge is identified with exactly one other edge and (b) the triangles identified at each vertex can always be arranged in a cycle $T_1, T_2, T_3, \ldots, T_k, T_1$ so that adjacent triangles are identified along an edge.*

Triangulable spaces are unquestionably the best understood of all topological spaces. This is due to their simple construction. The restriction to triangles is not essential, since every polygon can be triangulated (see Exercise 2). According to the definition, two triangles in a triangulable space have either two, one, or no vertexes in common, so that two triangles must have *different* sets of vertexes. It follows that the space can be described by labeling each vertex (identified vertexes being given the same label) and listing the triangles by vertex sets.

Example

Here is such a triangulable space consisting of eight triangles: *PQR*, *QRU*,
PRS, *RSU*, *PST*, *STU*, *PTQ*, *TQU*. Figure 20.3 shows how the space can be
partially assembled. We have first glued together the four triangles that are
identified at the vertex *P* (*PQR*, *PRS*, *PST*, *PTQ*) and then added the remaining triangles. The sides that remain to be identified are then labeled in
the usual way. If these remaining identifications are performed, the result
is a square pyramid. Thus this is a *surface* equivalent to a *sphere*.

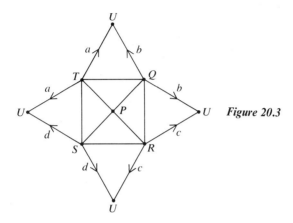

Figure 20.3

It is important now to find triangulations of the surfaces introduced in
§19 in order to demonstrate that they are all *surfaces* according to our formal
definition. It is not difficult to produce these triangulations, but a little care
is needed. Figure 20.4a shows a false triangulation of the torus. Although

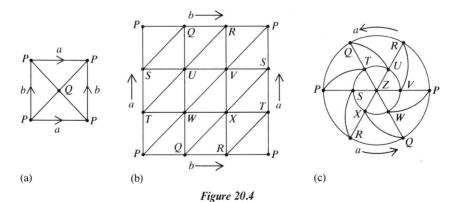

(a)   (b)   (c)

Figure 20.4

there are four triangles, there are only two vertexes and all four triangles have the same vertex set *PPQ*. Figure 20.4b gives a correct triangulation of the torus, and Figure 20.4c gives one for the projective plane. Since the connected sum of two surfaces is a surface (Exercise 3), these are the only triangulations we need to display.

Exercises

1. What is wrong with the following triangulations of (a) the torus and (b) the projective plane?

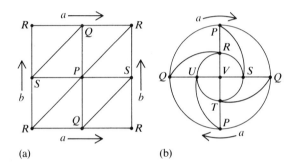

(a)   (b)

2. Define a topological space as **polygonable** if it can be obtained from polygons by the identification of edges and vertexes subject to the restriction that two polygons are identified either on an edge or at a single vertex or are completely disjoint. Prove that every polygonable space is triangulable.

3. Show that the connected sum of triangulable spaces is triangulable and that the connected sum of surfaces is a surface.

4. Show that the plane models of the torus and projective plane and their connected sums have the property that all vertexes are identified to a single vertex.

5. Exactly five of the following triangulable spaces are surfaces. Which?

(a) *PQS, QRS, PRS, PQT, QRT, PRT*

(b) *PQT, QRT, RST, PQU, QRU, PSU, PTU, PST*

(c) *PQR, QRS, RST, PST, PQT, TRU, QTU, QSU, PSU, PRU*

(d) *PQS, TUV, QRS, TVW, PQR, TUW, PRS, UVW*

(e) *PQR, PRS, PQS, PUV, PTU, PTV, SUV, QSV, QTV, QRT, RTU, RSU*

(f) *PQU, QRU, RSU, STU, PTU, PQV, QRV, RSV, STV, PTV*

(g) *PQS, QST, PRS, RSU, QRU, QTU, PQV, QVW, RWX, PRX,*
    QRW, PVX, TVW, TUW, UWX, SUX, SVX, STV

6. Let $\{P, Q, R, S\}$ be a set of four vertexes. Does the set of all triangles with vertexes from this set form a triangulable space? Is it a surface? If so, which one? Consider the same questions with regard to a set $\{P, Q, R, S, T\}$ of five vertexes.

Some surfaces may require an infinite number of triangles for their triangulation. Figure 20.5 gives an example. An infinite number of triangles have to be fitted around the one point that has been removed. The problem, as the following theorem shows, is that removing one point destroys the compactness of the disk.

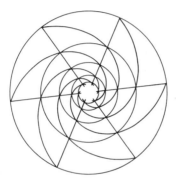

Figure 20.5 A disk
with one point removed.

Theorem 1

A surface is compact if and only if its triangulations use only a finite number of triangles.

This theorem gives a combinatorial interpretation of compactness. The next theorem gives a combinatorial interpretation of connectedness.

Theorem 2

A surface is connected if and only if the triangles in a triangulation of the surface can be arranged in a sequence $T_1, T_2, \ldots, T_n$ so that each triangle has at least one edge identified to an edge of an earlier triangle in the sequence.

Exercises

7. Prove Theorems 1 and 2.
8. Determine which of the surfaces in Exercise 5 are connected and which are compact.

§21 THE CLASSIFICATION THEOREM

This is the single most important theorem in the topological theory of surfaces.

Classification Theorem

Every compact, connected surface is topologically equivalent to a sphere, or a connected sum of tori, or a connected sum of projective planes.

In other words, assuming compactness and connectedness, the examples of surfaces given in §19 form a complete list of all surfaces. This is one of the most remarkable theorems in all topology. It is remarkable not only in its own right but in its isolation. There is no other theorem like it. The situation in higher dimensions appears, unfortunately, to be much more complicated than in two dimensions. No similar results are known for any other natural category of topological spaces.

The proof, although long, is thoroughly enjoyable. The theorem was first proved by Dehn and Heergaard (1907). Here we shall follow the classic exposition of Seifert and Threlfall [27].

To begin, let $\mathscr{S}$ be a compact, connected surface. The proof uses only two operations: (1) cutting a polygon into two pieces along an edge that is then labeled for later topological identification and (2) sewing together two edges that are labeled for identification. The definition of topological identification guarantees that these operations always produce topologically equivalent figures. Using these operations alone we shall assemble a model of $\mathscr{S}$ and transform this model into the model of a sphere, a connected sum of tori, or a connected sum of projective planes. In view of the classification theorem, these plane models are called the **normal forms** of the surfaces.

STEP 1. CONSTRUCTING A MODEL. By compactness, $\mathscr{S}$ has a triangulation using only a finite number of triangles. Furthermore, since $\mathscr{S}$ is connected, these triangles can be arranged in a sequence $T_1, T_2, \ldots, T_n$ so that each triangle is identified along an edge with an earlier triangle in the sequence. For example, let $\mathscr{S}$ be the surface {PQR, PST, RTV, PRV, PQT, TUV, PUV, QTU, WSU, PSU, QRS, RST}. This arrangement of the triangles is not satisfactory because, among other things, the third triangle RTV does not share an edge with either of the preceeding triangles. The triangles must be rearranged. There are many ways of managing this rearrangement. For example, we can begin by listing all the triangles in the cycle around the vertex P and then add the others as they seem to fit. The result is the sequence {PQR, PRV, PUV, PSU, PST, PQT, QTU, QRS, RTV, TUV, QSU, RST}.

We now take the first two triangles of the sequence and actually sew them together in the plane. The third triangle can then be sewn onto the figure formed by the first two, the fourth triangle can be sewn onto the figure formed by the first three, and so on. By construction, the next triangle always has an edge in common with the figure already formed in the plane. Because $\mathscr{S}$ is a surface, this edge in the figure is not identified with any other edge but the one in the next triangle. Thus the sewing can always be performed. At each stage the figure formed is topologically equivalent to a disk, as may be easily proven by induction. In this way we build a polygonal model of the surface $\mathscr{S}$. Figure 21.1a illustrates this process with the example of the preceeding paragraph.

The hypotheses of connectedness and compactness are both essential to this step. Connectedness guarantees that the next triangle can be glued to the figure already formed, while compactness guarantees that the whole process can be completed so that we can proceed to the next step of the proof.

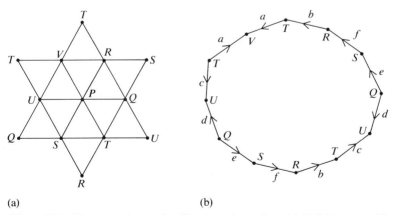

(a)   (b)

Figure 21.1   First steps in the classification of a surface. (a) Building a model. (b) Labeling the edges.

Having obtained this plane model of $\mathscr{S}$, however, we can forget topology. The remaining steps are purely combinatorial, designed to transform this model into a normal form recognizable as the form of one of the surfaces defined in §19.

On our polygonal model of $\mathscr{S}$ we can ignore the edges already identified. The remaining edges, on the boundary of the model, are still identified in pairs. We label these edges with letters and arrows to indicate this identification (see Figure 21.1b). The pairs of identified edges are of two types: those with arrows pointing in opposite directions around the polygon, and those with arrows pointing the same direction. The first are called **toroidal pairs**, the second **twisted pairs**. In our example there is only one toroidal pair (*a*).

STEP 2. ELIMINATION OF ADJACENT TOROIDAL PAIRS.   A toroidal pair of edges that adjoin can be eliminated directly by gluing. The justification for this step is given in Figure 21.2. Throughout the remainder of the proof, this step is to be repeated whenever any adjacent toroidal pairs occur. The only situation in which this step is *not* justified is when the toroidal pair is the *only* pair of edges left. Then the surface is a sphere and in this case the proof is concluded.

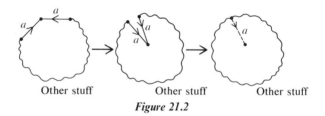

Other stuff   Other stuff   Other stuff

Figure 21.2

Exercises

1. Complete Step 2 for the example of Figure 21.1.
2. Carry out Steps 1 and 2 for the examples of surfaces given in Exercise 5 of §20. Don't forget that some of the examples in this exercise are not surfaces.
3. Carry out Steps 1 and 2 for these additional surfaces:

(h) *PQS, PRS, TUV, TVW, PRT, PTU, PQU, QUV, QSV, SVW, RSW, RTW*

(i) *QRS, RST, PST, PQT, RTU, QTU, QSU, PSU, PRU, QRV, RVW, PVW, PQW, RWX, QWX, QVX, PVX, PRX*

(j) *PQS, NQY, QYZ, PRX, QRW, PVX, TVW, TUW, TYZ, NTZ, QST, PRS, RSU, QRU, NQT, TUY, QUZ, PQV, RWX, UWX, NUZ, NUY, SUX, SVX, STV*

STEP 3. IDENTIFICATION OF ALL VERTEXES TO A SINGLE VERTEX.   At present the vertexes are divided into a number of different subsets: each subset consists of vertexes with the same label, all these vertexes being identified with each other. In the example of Figure 21.1b there are five such sets corresponding to the letters Q, R, S, T, and U, each consisting of two vertexes (V was eliminated in Step 2!). In this step we replace this model with one in which all vertexes are identified to a single vertex and therefore all vertexes have the same label. Here is how: if all vertexes are not already labeled the same, then somewhere there is an adjacent pair of vertexes with different labels. Let these labels be P and Q (see Figure 21.3). The two edges meeting at P cannot be identified, for if we had ⟨figure⟩ ,then P and Q would be identified contrary to our assumption, while the configuration ⟨figure⟩ was eliminated in Step 2. Thus the configuration of edges at P must be as shown in Figure 21.3: the edges a and b are not identified. Somewhere else on the model there is another edge b, which naturally contains another vertex labeled P. Making the cut c and gluing the edges b together as indicated in Figure 21.3, we obtain a model with one less vertex labeled P and one more vertex labeled Q. It is clear that by repeating this process, stopping to repeat Step 2 as necessary, we can systematically increase the number of Q vertexes at the expense of all the other types. Eventually all vertexes are identified together.

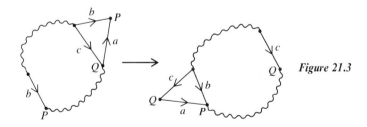

Figure 21.3

STEP 4. NORMALIZATION OF TWISTED PAIRS.   The purpose of this step is to transform the model so that in all the twisted pairs the two edges will adjoin each other. Supposing the model contains a twisted pair that is not

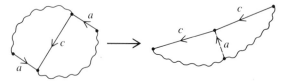

Figure 21.4

adjacent, the cutting and pasting operation depicted in Figure 21.4 will bring them together. This operation must be repeated until all the twisted pairs adjoin. At this point all the toroidal pairs may vanish under the repetitions of Step 2. For such surfaces the proof is concluded: they are connected sums of projective planes.

STEP 5. NORMALIZATION OF TOROIDAL PAIRS. Suppose the model contains a toroidal pair. Then we claim that there is a second toroidal pair separating the edges of the first pair from each other; that is, the two pairs alternate going around the polygonal model. To see this, suppose there were *no* such separating pair. The situation is depicted in Figure 21.5. The portions of the polygon A and B, separating the toroidal pair cc, each contain whole pairs of edges since all the twisted pairs adjoin, thanks to Step 4, while by assumption there is no separating toroidal pair. Now, if A contains whole

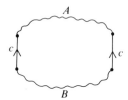

Figure 21.5

pairs of edges only, then the vertexes of A are identified only with other vertexes of A. In other words, it follows that no vertex of A is identified to any vertex of B. This contradicts Step 3. Therefore there must be a separating toroidal pair dd. Then the sequence of cutting and pasting operations shown in Figure 21.6 will bring the four edges together. These operations are repeated until all the toroidal pairs are arranged in adjoining interleaved pairs of pairs. At this point it may be that there are no twisted pairs. For such surfaces the proof is concluded; they are connected sums of tori.

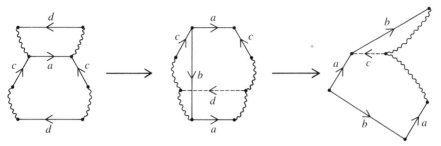

Figure 21.6

STEP 6. TRANSFORMATION OF TOROIDAL PAIRS INTO TWISTED PAIRS.
We now have a model containing both types of pairs normalized according
to the preceeding two steps. The sequence of cutting and pasting operations
displayed in Figure 21.7 shows how given one twisted pair and two normalized toroidal pairs, the toroidal pairs can be converted to twisted pairs.
These operations must be repeated until all the toroidal pairs are transformed
to twisted pairs. The surface is therefore a connected sum of projective planes.
Thus every compact, connected surface is either a sphere (Step 2), a connected sum of tori (Step 5), or a connected sum of projective planes (Step 4
or Step 6). This concludes the proof.

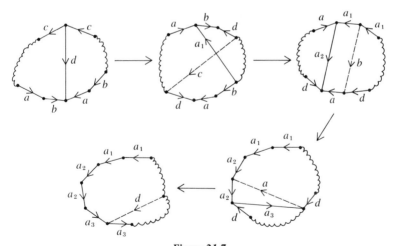

Figure 21.7

To the extent that the goal of topology is to understand the variety of
possible topological spaces, this theorem represents an ideal. The compact,

connected surfaces are a natural class of topological spaces; their structure is completely described by the classification theorem. Unfortunately, despite intensive attacks, other classes of topological spaces continue to resist classification.

Exercises

4. Complete the classification of the surfaces begun in Exercises 1, 2, and 3. Keep your eye out for short cuts!

5. How is the connected sum of a torus and a Klein bottle classified? What about the connected sum of two Klein bottles?

6. Show that it is possible by cutting and pasting operations to convert a toroidal pair of edges into a twisted pair of edges and vice versa. Show that this cannot affect the proof of the classification theorem, since once a pair of edges is normalized, no subsequent step will alter its type.

7. State and prove a classification theorem for compact but not necessarily connected surfaces.

8. Prove that a connected sum of projective planes is topologically equivalent to a connected sum of tori in connected sum with either a projective plane or a Klein bottle.

9. (Seifert and Threlfall) Show that every compact, connected surface is topologically equivalent to a surface of the following type, where each of the n pairs of edges is toroidal except the last, which can be either toroidal or twisted, depending on the surface.

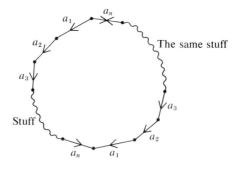

§22 SURFACES WITH BOUNDARY

The classification theorem for compact surfaces does not extend in any way to noncompact surfaces. There is an incredible variety of noncompact surfaces, in part because every open subset of a surface is itself a surface. The classification theorem can be extended by considering compact surfaces with boundary. Intuitively, a surface with boundary is a topological space in which every point *either* has a neighborhood equivalent to a disk (as on a surface) *or* has a neighborhood equivalent to a half disk. Typical examples of surfaces with boundary are the cylinder and the Möbius strip. Here is the combinatorial definition.

Definition

A **surface with boundary** *is a topological space obtained by identifying edges and vertexes of a set of triangles according to all the requirements of the definition of surface except that certain edges may not be identified with another edge. These edges, which violate the definition of a surface, are called* **boundary edges,** *and their vertexes, which also violate the definition of surface, are called* **boundary vertexes.**

The cylinder is equivalent to a sphere with two disks cut out (Exercise 1 of §19), while the Möbius strip is equivalent to a projective plane with one hole (Exercise 6 of §19). These facts suggest the following theorem.

Classification Theorem for Surfaces with Boundary

Every compact, connected surface with boundary is equivalent to either a sphere or a connected sum of tori or a connected sum of projective planes, in any case with some (finite) number of disks removed.

Let $\mathscr{S}$ be a surface with boundary. The idea of the proof is first to assemble the holes and then follow the steps of the proof of the classification theorem for surfaces. The first step will be to construct a polygonal model of $\mathscr{S}$ in which all the edges are identified in pairs, only this time the polygon will have holes. Before beginning, we require some further assumptions regarding the triangles making up the surface $\mathscr{S}$. These are that (1) no triangle has more than one boundary edge, and (2) no edge not in the boundary can have both vertexes in the boundary. It may well be that $\mathscr{S}$, as initially triangulated, does not satisfy these conditions. We then subdivide the

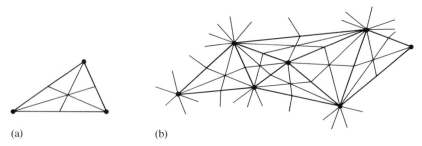

Figure 22.1   Barycentric subdivision (a) of a triangle and (b) of a surface.

surface according to the scheme depicted in Figure 22.1 (**barycentric sub-division**). You will easily verify that the triangulation resulting from a subdivision must satisfy both (1) and (2).

Let us call a triangle that intersects the boundary a **boundary triangle**, whether this intersection is an edge or just a vertex. It follows from (1) that every boundary triangle has exactly two edges that while not on the boundary have one vertex on the boundary. On the other hand, every edge that is not on the boundary but that has a vertex on the boundary lies on exactly two boundary triangles. Now starting with a given boundary triangle T_1 we can find an edge e_1 on T_1, not on the boundary but containing a boundary vertex. The edge e_1 lies on a second boundary triangle T_2. The triangle T_2 in turn contains a second edge e_2 not on the boundary but with a boundary vertex. The edge e_2 lies on another boundary triangle T_3, and so on. In this manner we obtain a sequence of alternating triangles and edges, $T_1 e_1 T_2 e_2 \cdots T_n$, each edge belonging to the two adjacent triangles, and each triangle containing the two adjacent edges. Such a sequence must eventually return to the first triangle closing itself off, that is, $T_n = T_1$. We now glue these triangles together along these edges one by one through the sequence. As each triangle is added, the figure stays topologically equivalent to a disk. Finally T_{n-1} is glued to T_1 along the edge e_{n-1}. Since this amounts to gluing two edges of the same polygon together, the result is an annulus, topologically equivalent to a disk with a hole. We can place this annulus in the plane with the boundary edges on the inside so that the outside edges all remain to be identified to other triangles of the surface. Figure 22.2 gives an example of a cycle of sixteen triangles. There is no possibility of a boundary edge appearing on the outside of the annulus, because each of the edges e_i has only one vertex on the boundary.

The construction of the preceeding paragraph is now repeated until all the boundary triangles are incorporated in annuli of this type. Then the annuli, together with the remaining triangles of $\mathscr{S}$, can be assembled into a single polygonal model of the surface in which the boundary appears as a

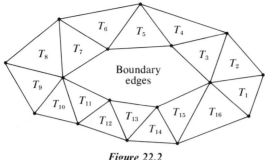

Figure 22.2

series of holes in the polygon. The outside edges of this model are identified in pairs, just as in the proof of the classification theorem. The remaining steps of the proof of that theorem can now be applied directly to this model, care being taken to avoid the holes when cutting the model. This completes the proof.

Exercises

1. Classify the following surfaces with boundary. To help you get started, note that the edge ST is in the boundary of each example.

(a) $PQU, PTU, QRV, RVW, PRW, PQX, PWX, QRS, QSX, PRT, RST$

(b) $PQV, QVW, QRW, RSW, SWX, STX, VWX, PVX, TUX$

(c) $PSW, PSU, PUV, RUV, RVX, RXY, QXY, QTY, QTW, STW$

2. Justify the statement made in the text concerning barycentric subdivision.

3. Explain why the sequence $T_1 e_1 T_2 e_2 \cdots$ must turn into a cycle.

Notes. All discussions of the classification theorem go back to Seifert and Threlfall [27]. You can see a slightly different treatment in the books by Blackett [4] and Frechet and Fan [8].

five

Homology of Complexes

§23 COMPLEXES

Homology provides the tool that will enable us to tackle simultaneously a variety of topological problems. Although we will concentrate on surfaces, homology is applicable to a large class of other topological spaces of a similar combinatorial nature. The purpose of this section is to introduce these spaces, called complexes.

Definition

A **complex** *is any topological space that is constructed out of vertexes, edges, and polygons by topological identification.*

It is important to emphasize that according to the definition a complex is not simply a topological space, but a topological space together with information on how to construct that space by gluing together vertexes, edges, and polygons. A complex is a topological space plus a combinatorial structure that describes how the space is created. Thus the same topological space can give rise to many different complexes. The sphere, for example, has many distinct triangulations, each of which makes the sphere into a distinct complex. Complexes include all the types of combinatorial structures that have appeared in this book so far: *polyhedra* (Chapter One), *triangulations of plane sets* (Chapter Two), *gratings* (Chapter Three), and *triangulations of surfaces* (Chapter Four). In addition, the polygons in complexes may be *self-identified* along a pair of edges, so that complexes include the *normal forms of surfaces* developed in Chapter Four. Finally there are a number of other possible types of identification that might occur in a complex. Figure 23.1 gives some idea of these possibilities. Parts (a) and (b) will be familiar, but we also have (c) identification of polygons at vertexes without identification of the corresponding edges, (d) identification of edges at vertexes without polygons at all, and (e) a combination of ingredients. We are content to present these examples without defining precisely the limits of topological identification, because most applications only involve triangulations, which were defined precisely in §20. The idea of a complex is due to Poincaré.

As in Chapter Three, it will be convenient to adopt a systematic terminology for vertexes, edges, and polygons. Once again they will all be called simplexes. Thus *0-simplex* will be synonymous with point and vertex, *1-simplex* will be synonymous with edge, and *2-simplex* will be synonymous with face and polygon. Homology is a means for treating algebraically certain relationships among the simplexes of a complex.

The most important of these relationships is the boundary relationship. You will recall how the homology theory of Chapter Three depended

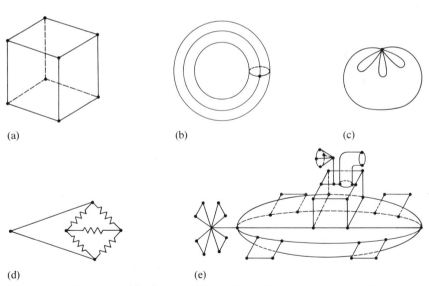

(a)   (b)   (c)

(d)   (e)

Figure 23.1 Some complexes. (a) A sphere (6 faces, 12 edges, 8 vertexes). (b) A torus (1 face, 2 edges, 1 vertex). (c) A bandana (1 face, 4 edges, 1 vertex). (d) An electrical circuit (no faces, 7 edges, 5 vertexes). (e) *Das Unterseeboot.*

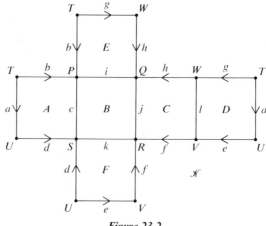

Figure 23.2

entirely on the notion of boundary embodied in the boundary operator. The homology of this chapter imitates that of Chapter Three closely in this respect. The boundary of a simplex *by itself* is clear: the boundary of a polygon is the set of its edges, the boundary of an edge is the set of its end-points, while a point has no boundary. However, the boundary of a simplex *in a complex* requires more careful treatment. A few examples will show why.

Examples

1. We first choose an example in which nothing unusual occurs. Let $\mathcal{K}$ be the complex equivalent to a sphere depicted in Figure 23.1a. In Figure 23.2 $\mathcal{K}$ has been cut open and unrolled into the plane with suitable identifications marked. Every simplex of $\mathcal{K}$ has been carefully labeled. We see that $\mathcal{K}$ consists of vertexes P, Q, R, S, T, U, V, W; edges $a, b, c, d, e, f, g, h, i, j, k, l$; and faces A, B, C, D, E, F. The tables in Figure 23.3 summarize the boundary relation on $\mathcal{K}$. Table (a) gives the boundaries of the edges, while table (b) gives the boundaries of the faces of $\mathcal{K}$. In these tables an entry of one (1) means "yes, on the boundary," while an entry of zero (0) means "no, not on the boundary." You should verify that these tables do reflect the actual boundaries as depicted in Figure 23.2. The entries in these tables are called the **incidence coefficients of** $\mathcal{K}$.

2. Next consider the complex $\mathcal{L}$ equivalent to the torus given in Figure 23.1b. Unrolled, as in Figure 23.4, this is the familiar normal form of the torus consisting of one vertex, P, two edges, a and b, and one face, A. Incidence coefficients, given in Figure 23.3c and d reveal a new phenomenon. The entry 2 in the first row of the first table means that the vertex P appears

(a)

	P	Q	R	S	T	U	V	W
a	0	0	0	0	1	1	0	0
b	1	0	0	0	1	0	0	0
c	1	0	0	1	0	0	0	0
d	0	0	0	1	0	1	0	0
e	0	0	0	0	0	1	1	0
f	0	0	1	0	0	0	1	0
g	0	0	0	0	1	0	0	1
h	0	1	0	0	0	0	0	1
i	1	1	0	0	0	0	0	0
j	0	1	1	0	0	0	0	0
k	0	0	1	1	0	0	0	0
l	0	0	0	0	0	0	1	1

for $\mathscr{X}$ (b)

	a	b	c	d	e	f	g	h	i	j	k	l
A	1	1	1	1	0	0	0	0	0	0	0	0
B	0	0	1	0	0	0	0	0	1	1	1	0
C	0	0	0	0	0	1	0	1	0	1	0	1
D	1	0	0	0	1	0	1	0	0	0	0	1
E	0	1	0	0	0	0	1	1	1	0	0	0
F	0	0	0	1	1	1	0	0	0	0	1	0

(c)

	P
a	2
b	2

for $\mathscr{L}$ (d)

	a	b
A	2	2

Figure 23.3   Incidence coefficients.

twice in the boundary of the edge, as is clearly the case in Figure 23.4. Similarly, the first 2 in the second table means that the edge a appears twice in the boundary of the polygon A. The polygon A is a rectangle, just as the polygon A in Figure 23.2 is a rectangle. Both have four edges. In Figure 23.2, however, there are four *different* edges, while in Figure 23.4 two edges appear twice, reflecting the self-identification of the rectangle. The incidence coefficients in turn reflect this circumstance. The incidence coefficients are chosen to convey as much information as possible about the boundaries of the simplexes of $\mathscr{L}$ *in the complex* $\mathscr{L}$. Here is the formal definition of incidence coefficients.

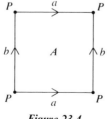

Figure 23.4

Definition

*Let $\mathcal{K}$ be a complex. Let x be a k-simplex and y be a $(k+1)$-simplex of $\mathcal{K}$ ($k = 0, 1$). The **incidence coefficient** of x in y is the number of times that x appears in the boundary of y.*

The incidence coefficients contain a good deal of information about the complex. That is good because the definition of the homology groups depends entirely on the incidence coefficients. However, the incidence coefficients alone do *not* determine the complex completely. It would be impossible, for example, to construct the torus from the information given in Tables (c) and (d) of Figure 23.3. One would know that the polygon A was a rectangle with two sides a and two sides b but would not be able to tell what their proper order was on the boundary of A, or whether they were twisted or not. That is unfortunate because if this information is missing from the incidence coefficients, it will also be missing from the homology groups. Thus the homology groups, as we shall see, contain only partial information about the topological space determined by the complex $\mathcal{K}$.

Exercises

1. For the following examples of complexes, label all simplexes carefully, taking identifications into account, and construct the tables of incidence coefficients.

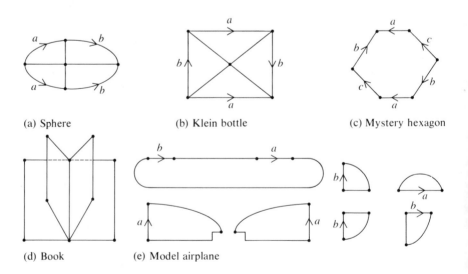

(a) Sphere   (b) Klein bottle   (c) Mystery hexagon

(d) Book   (e) Model airplane

2. The surface shown below is called a **hole through a hole in a hole**. Show that this is a complex by finding a means for constructing it from polygons.

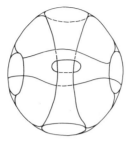

3. Prove that a complex $\mathcal{K}$ is connected if and only if any two vertexes can be connected by a sequence of edges each of which shares an endpoint with its neighbors in the sequence.

GRAPH THEORY

A complex consisting of vertexes and edges without faces is called a **graph**. An example was given in Figure 23.1d. Figure 23.5 gives some further examples. We are interested in the topological structure of graphs. Therefore the shape of the edges is not important. All that matters is that certain pairs

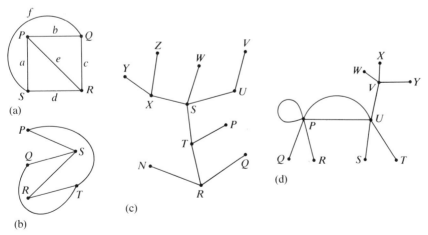

Figure 23.5 Some graphs. (a) K_4. (b) $K_{3,2}$. (c) A tree. (d) Miscellaneous.

of vertexes are connected by edges, while other pairs of vertexes are not. A vertex may be connected to itself by an edge, which is then called a **loop**, and there may be several edges between two vertexes. Both phenomena occur in Figure 23.5d. Graphs have many applications, to electric circuits and transportation networks, for example. Here we study the small portion of graph theory that is described by the phrase "one-dimensional combinatorial topology."

Definition

A **walk** *on a graph is an alternating sequence of vertexes and edges in which each edge is flanked by its own vertexes, no edge is used more than once, but at least one edge is used. For example, PbQfSdR is a walk on Figure 23.5a. A* **cycle** *is a walk that ends with the vertex with which it began.*

Examples of Graphs

1. *Complete graphs.* The **complete graph on *n* vertexes**, abbreviated K_n, is the graph with n vertexes and exactly one edge between every distinct pair of vertexes. Figure 23.5a is the complete graph on four vertexes.

The **complete bipartite graph**, abbreviated $K_{n,m}$, is the graph on $n + m$ vertexes divided into two subsets, one of n vertexes and the other of m vertexes, with no edges connecting vertexes of the same subset and exactly one edge connecting each vertex of one subset with each vertex of the other subset (Figure 23.5b).

2. *Connected graphs.* A graph is **connected** if there is at least one walk between any two vertexes. All the examples in Figure 23.5 are connected. In general, a graph will be the union of one or more connected subgraphs, called **components**. Figure 23.5 as a whole could be regarded as a single graph with four components. This terminology agrees with our earlier notion of connectedness.

3. *Trees.* A **tree** is a connected graph without cycles. Figure 23.6 gives all the trees with six vertexes. Trees play an important role in the topology

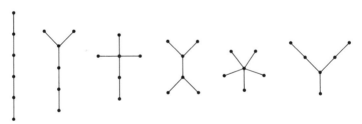

Figure 23.6 Some trees.

of graphs. Some of their special properties are described in the following theorem.

Theorem

Let $\mathscr{G}$ be a tree. Then

(a) *Every pair of vertexes in $\mathscr{G}$ is connected by a unique walk.*

(b) *The graph obtained by removing any one edge from $\mathscr{G}$ is disconnected.*

(c) *$\mathscr{G}$ has one more vertex than edge: $V = E + 1$, where V and E are the numbers of vertexes and edges, respectively, in $\mathscr{G}$.*

To prove (a), let P and Q be vertexes of $\mathscr{G}$, and suppose, contrary to what we want to prove, that P and Q are joined by two different walks λ and μ. If λ and μ consist of completely different edges, then λ and μ together form a cycle beginning at P, going to Q via λ, and returning to P via μ. Since $\mathscr{G}$ has no cycles, this is impossible, so λ and μ must have some edges in common. Nevertheless we can deduce a contradiction and therefore prove (a) in this way: let R be the first vertex along the path λ that is on an edge e not also in μ. Thus, for example, if λ and μ begin with *different* edges, then $R = P$. The vertex R will be in both λ and μ but λ and μ will use different edges issuing from R, so that the next vertexes in λ and μ will be different. Such a vertex R must exist because λ and μ are different by assumption. Let T be the next vertex that is again in both λ and μ. At the worst $T = Q$. Then by construction R and T are joined by two completely different walks, contradicting the fact that $\mathscr{G}$ has no cycles.

To prove (b), consider a single edge e of $\mathscr{G}$ connecting the points P and Q. According to (a), P and Q are connected by a *unique* walk. This walk must be the walk PeQ. Removing e from $\mathscr{G}$, therefore, produces a graph in which the vertexes P and Q, at least, cannot be connected, that is, a disconnected graph. This proves (b).

The proof of (c) is by induction on the number of edges E in the graph $\mathscr{G}$. If $\mathscr{G}$ has just one vertex, then $V = 1, E = 0$, and obviously $V = E + 1$. Suppose the theorem is true for all trees of k or fewer edges. We shall prove that it is true for a graph $\mathscr{G}$ with $(k + 1)$ edges. Let e be an edge of $\mathscr{G}$. According to (b), by removing the edge e, $\mathscr{G}$ is disconnected into two components $\mathscr{G}_1$ and $\mathscr{G}_2$ whose union, together with the edge e, is $\mathscr{G}$. Let V_1 (V_2) be the number of vertexes in $\mathscr{G}_1$ ($\mathscr{G}_2$), and let E_1 (E_2) be the number of edges of $\mathscr{G}_1$ ($\mathscr{G}_2$). Then $V = V_1 + V_2$ while $E = E_1 + E_2 + 1$. Since both $\mathscr{G}_1$ and $\mathscr{G}_2$ are trees of k or fewer edges, by the induction assumption $V_1 = E_1 + 1$ and $V_2 = E_2 + 1$. Thus $V = V_1 + V_2 = E_1 + 1 + E_2 + 1 = E + 1$. This completes the proof of (c).

Exercises

4. Let $\mathscr{G}$ be a graph and let V be a vertex of $\mathscr{G}$. The number of edges meeting at V (loops count twice) is called the **degree** of V. If V has degree one, V is called an **endpoint** of $\mathscr{G}$. Find examples of graphs with no endpoints and with one endpoint. Find a graph that is all endpoints.

5. *The Seven Bridges of Königsberg.* The city of Königsberg in East Prussia (now Kaliningrad R.S.F.S.R.) lies on the river Pregel. Two islands in the river are connected with each other and the shore by seven bridges, as shown below.

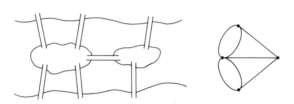

The problem faced by the citizens of Königsberg was to find a walk that crosses each bridge exactly once. Euler solved this problem in 1736 by abstracting from the map the graph shown to its right. The problem now is to find a walk in the technical sense that uses every edge. Euler proved that no such walk exists. More generally he proved that a graph $\mathscr{G}$ has a walk passing through every edge if and only if at most two vertexes have odd degree. Prove this result.

6. The table below gives the number of topologically distinct trees with up to eight vertexes. Find these trees. Be careful, because two trees can look very different yet be topologically equivalent.

Number of vertexes	1	2	3	4	5	6	7	8
Number of trees	1	1	1	2	3	6	11	23

7. Let $\mathscr{G}$ be a graph. Prove the following converses to the theorem on trees:

(a) If any two vertexes of $\mathscr{G}$ are joined by a unique walk, then $\mathscr{G}$ is a tree.

(b) If $\mathscr{G}$ is connected but becomes disconnected by the removal of any one edge, then $\mathscr{G}$ is a tree.

(c) If $V = E + 1$ and $\mathscr{G}$ is connected, then $\mathscr{G}$ is a tree.

8. Let $\mathcal{T}$ be a tree with vertexes $P_1, P_2, \ldots, P_V$. Let $d_1, d_2, \ldots, d_V$ be the corresponding degrees. Show that

$$d_1 + d_2 + \cdots + d_V = 2(V - 1)$$

Conclude that every tree has at least two endpoints.

9. Let $\mathcal{G}$ be a graph. A **subtree** of $\mathcal{G}$ is a tree made up from edges and vertexes of $\mathcal{G}$. A **spanning tree** is a subtree that uses all the vertexes of $\mathcal{G}$. If V is the number of vertexes of $\mathcal{G}$, show that a subtree $\mathcal{T}$ is a spanning tree if and only if $\mathcal{T}$ has $V - 1$ edges.

THREE-DIMENSIONAL COMPLEXES

The complexes introduced so far have been two-dimensional (surfaces, for example) or one-dimensional (graphs). You can easily imagine zero-dimensional complexes (discrete sets of points). The idea of complex extends likewise to higher dimensions. Let us illustrate this in three dimensions. The building blocks for three-dimensional complexes are points, edges, polygons, and polyhedral solids. A **polyhedral solid** is any subset of three-dimensional space that is topologically equivalent to a spherical ball and whose boundary surface has been divided into faces, edges, and vertexes (i.e., the boundary surface is a polyhedron). The polyhedral solids are the natural generalization to three dimensions of polygons. A **three-dimensional complex** is a topological space made up of polyhedral solids glued together by topological identification along faces. Without going into detail, note that identification of poly-hedral solids along faces is more complicated than identification of polygons along edges; for while any two polygons can be identified along any edge, two polyhedral faces must be identified along faces with the same number of edges. An example of a three-dimensional complex is given in Figure 23.7. It consists of a single cube with opposite faces identified in pairs. Each point on each face is identified with the point directly opposite on the parallel face, for example the point Q in Figure 23.7. Note that this identification of faces forces the identification of edges in sets of *four* while all *eight* vertexes are identified to one. Every point in this complex has a neighborhood equivalent to a solid sphere. In Figure 23.7, for example, the neighborhoods of the point Q consist of two half spheres that together form a whole solid sphere. Three-dimensional complexes with this property are called **three-dimensional manifolds**. By analogy, surfaces are frequently called **two-dimensional manifolds**. It is an outstanding unsolved problem to classify three-dimensional

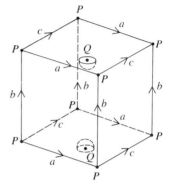

Figure 23.7 The three-dimensional torus.

manifolds. This problem is considered extremely difficult. There is a much greater variety of three-dimensional manifolds than two. No theorem remotely like the classification theorem of Chapter Four has been found for three-dimensional manifolds.

Exercises

10. Draw neighborhoods at the edges and the vertex of the three-dimensional torus, and verify that these points have neighborhoods equivalent to a solid sphere.

11. Consider the problem of generalizing the idea of triangulation from two to three dimensions. What solid should play the role in three dimensions that the triangle plays in two? Decide what rules, analogous to those given for triangulations (§20), should govern the identification of these solids.

12. Consider the three-dimensional complex obtained by identifying opposite sides of each of the following solids. In each case determine the pattern of identification of edges and vertexes imposed by the identification of faces. Is either of these complexes a three-dimensional manifold?

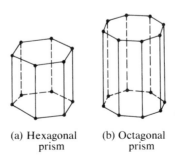

(a) Hexagonal   (b) Octagonal
    prism   prism

13. Formulate an appropriate definition of one-dimensional manifolds, and find examples of complexes satisfying your definition. What about zero-dimensional manifolds?

---

§24 HOMOLOGY GROUPS OF A COMPLEX

We begin this section with some background material on groups.

Definition

*An Abelian **group** is a set G together with a binary operation $(+)$ satisfying the following axioms:*

(a) **associativity**: *For x, y, z in G, $(x + y) + z = x + (y + z)$.*

(b) **commutativity**: *For x, y in G, $x + y = y + x$.*

(c) **zero**: *There is a unique element 0 in G such that $x + 0 = x$ for all x in G.*

(d) **inverses**: *For each x in G there is an element $(-x)$ of G such that $x + (-x) = 0$.*

The term "Abelian" refers to the commutative axiom, which is not normally part of the definition of group. We will usually omit the word "Abelian," since *all* the groups arising in homology are Abelian; however, we give one example of a noncommutative group $\mathscr{D}_3$ (below).

Examples

1. *Small groups.* Groups with only a few elements can be given by an addition table. Below we present all the groups of six elements or less. For the numbers 1, 2, 3, and 5 there is just one group of that many elements. Those groups as it happens are **cyclic**, meaning that the whole group can be obtained by repeatedly adding a certain element of the group to itself. The elements with this property are called **generators** of the group. For example, c is a generator for the group $\mathscr{C}_5$, since we have $c + c = a, c + a = d, c + d = b, c + b = 0$. There are two groups of order four, a cyclic group and another, the Klein four group, that plays an interesting role in topology, as we shall see. There are also two groups of order six. Note that one of these, the dihedral group, is not commutative. For example, we have in $\mathscr{D}_3$, $a + b = e$ but $b + a = c$.

+	0
0	0

The trivial
group, $\mathscr{C}_1$

+	0	a
0	0	a
a	a	0

The cyclic group
of order two, $\mathscr{C}_2$

+	0	a	b
0	0	a	b
a	a	b	0
b	b	0	a

The cyclic group
of order three, $\mathscr{C}_3$

+	0	a	b	c
0	0	a	b	c
a	a	0	c	b
b	b	c	0	a
c	c	b	a	0

Klein's four group,
also called the
dihedral group, $\mathscr{D}_2$

+	0	a	b	c
0	0	a	b	c
a	a	b	c	0
b	b	c	0	a
c	c	0	a	b

The cyclic group
of order four, $\mathscr{C}_4$

+	0	a	b	c	d
0	0	a	b	c	d
a	a	b	c	d	0
b	b	c	d	0	a
c	c	d	0	a	b
d	d	0	a	b	c

The cyclic group
of order five, $\mathscr{C}_5$

+	0	a	b	c	d	e
0	0	a	b	c	d	e
a	a	0	e	d	c	b
b	b	c	d	e	0	a
c	c	b	a	0	e	d
d	d	e	0	a	b	c
e	e	d	c	b	a	0

The dihedral group
of order six, $\mathscr{D}_3$

+	0	a	b	c	d	e
0	0	a	b	c	d	e
a	a	b	c	d	e	0
b	b	c	d	e	0	a
c	c	d	e	0	a	b
d	d	e	0	a	b	c
e	e	0	a	b	c	d

The cyclic group
of order six, $\mathscr{C}_6$

2. *Infinite groups.* The most important of these in homology is the group $\mathscr{Z}$ of integers, $\mathscr{Z} = \{\ldots, -2, -1, 0, 1, 2, \ldots\}$. Other types of "numbers" provide further examples of infinite groups. Thus the rational numbers, the real numbers, and the complex numbers are also groups (under addition). In a different direction the set $\mathscr{Z}^2$ of all ordered pairs (a, b) of integers a, $b \in \mathscr{Z}$ forms a group. The elements in this group may be regarded as the points in the Cartesian plane with integer coordinates (Figure 24.1). The group operation is the same as vector addition.

3. *Symmetry groups.* These groups play an important role in geometry. Given a geometric figure, its **symmetry group** is the set of all rotations and reflections that preserve the appearance of the figure. Consider Figure 24.2a, for example. This figure remains the same under rotations of $0°$, $90°$, $180°$, and $270°$. Let these transformations be represented by the letters R_0, R_{90}, R_{180}, R_{270}. Of course R_0 is the identity transformation. Using composition of transformations as the group operation, these four transformations already

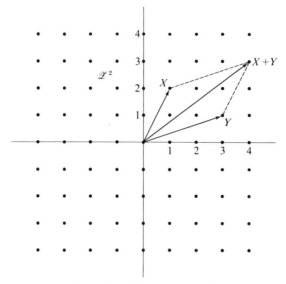

Figure 24.1   The group $\mathscr{L}^2$.

form a group, the **rotation group** of the figure. The addition table for this group is given in Figure 24.2b. Since the figure has no reflections, this is the whole symmetry group. On the other hand, the appearance of Figure 24.3a is preserved not only by the rotations R_0 and R_{180} (but *not* by R_{90} or R_{270}) but also by a horizontal reflection H and a vertical reflection V. The addition table for this symmetry group appears in Figure 24.3b.

Often it is important to decide whether two groups have the same algebraic structure. The following definition establishes a precise criterion.

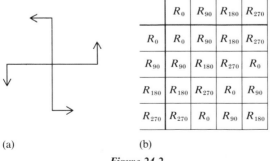

	R_0	R_{90}	R_{180}	R_{270}
R_0	R_0	R_{90}	R_{180}	R_{270}
R_{90}	R_{90}	R_{180}	R_{270}	R_0
R_{180}	R_{180}	R_{270}	R_0	R_{90}
R_{270}	R_{270}	R_0	R_{90}	R_{180}

(a)   (b)

Figure 24.2

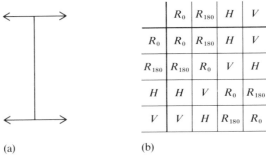

	R_0	R_{180}	H	V
R_0	R_0	R_{180}	H	V
R_{180}	R_{180}	R_0	V	H
H	H	V	R_0	R_{180}
V	V	H	R_{180}	R_0

(a)   (b)

Figure 24.3

Definition

Let G_1 and G_2 be groups. A **homomorphism** *is a function f from G_1 to G_2 that is additive, meaning for any x and y in G_1, $f(x + y) = f(x) + f(y)$. An* **isomorphism** *is a homomorphism that has an inverse homomorphism. Two groups are* **algebraically equivalent** *or* **isomorphic** *when there is an isomorphism between them. This is symbolized by writing $G_1 \cong G_2$.*

For example, the symmetry group of Figure 24.2 is algebraically equivalent to the cyclic group of order 4, an isomorphism being $f(R_0) = 0$, $f(R_{90}) = a$, $f(R_{180}) = b$, $f(R_{270}) = c$; the symmetry group of Figure 24.3 is isomorphic to the Klein four group via the isomorphism $f(R_0) = 0$, $f(R_{180}) = a$, $f(H) = b$, $f(V) = c$.

Exercises

1. Find all the generators for each of the cyclic groups given in the text.
2. Describe the symmetry group, including the addition table, of each of the following figures. In each case find a small group isomorphic to the symmetry group.

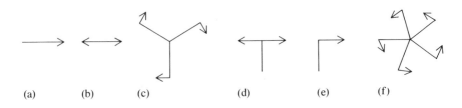

(a)   (b)   (c)   (d)   (e)   (f)

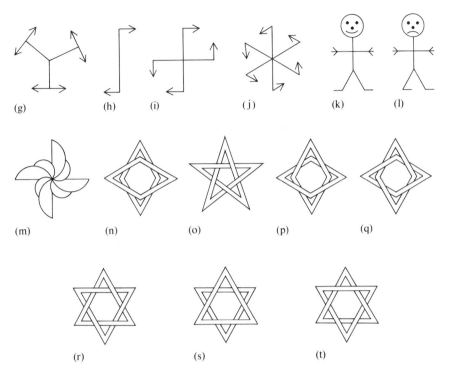

(g)   (h)   (i)   (j)   (k)   (l)

(m)   (n)   (o)   (p)   (q)

(r)   (s)   (t)

3. Find the symmetry groups of the following familiar geometric figures: a rectangle, a rhombus, a general parallelogram, a scalene triangle, an isosceles triangle, and an equilateral triangle.

4. Find a second isomorphism of the symmetry group of Figure 24.2 with $\mathscr{C}_4$. Do the same with symmetry group of Figure 24.3 and $\mathscr{D}_2$.

5. Let G_1 and G_2 be groups and $f : G_1 \to G_2$ be a group homomorphism. Show that $f(0) = 0$ and $f(-x) = -f(x)$.

6. A group is called an **idemgroup** when $x + x = 0$ for all elements of the group; in other words, $x = -x$ for all elements of the group. Which of the groups introduced in this section are idemgroups?

Let $\mathscr{K}$ be a complex. The definition of the homology groups for $\mathscr{K}$ parallels very closely the definitions given in Chapter Three for the special case of gratings. On this account most of the proofs will be left to you. First come the chains and the chain groups.

Definition

*A **k-chain** of $\mathscr{K}$ is a set of k-simplexes $(k = 0, 1, 2)$. The sum $C_1 + C_2$ of two k-chains C_1 and C_2 is defined to be the set of k-simplexes contained in C_1 or C_2 but not contained in both. (In set theory the technical name for this operation is symmetric difference.) With this operation, the set $C_k(\mathscr{K})$ of k-chains is a group in which the identity is the empty set $\varnothing$ of k-simplexes. In addition, each k-chain is its own inverse so that $C_k(\mathscr{K})$ is an idemgroup.*

For example, consider the complex $\mathscr{K}$ described in Figure 23.2. Typical k-chains are $P + Q + V$ $(k = 0)$, $a + c + f + g$ $(k = 1)$ and $B + C + D$ $(k = 2)$. From the definition it is clear that the algebra of the chains all by itself can tell us nothing about the topology of the complex $\mathscr{K}$, since the operation of chain addition is purely a set operation and has no connection with the geometry of $\mathscr{K}$. The connection with geometry is provided by the boundary operator.

Definition

*Let C be a k-chain of $\mathscr{K}$. The **boundary of C**, $\partial(C)$, is the $(k - 1)$-chain consisting of those $(k - 1)$-simplexes that are incident a total of an odd number of times with the simplexes of C, where incidence is determined by the incidence coefficients of $\mathscr{K}$. The boundary operator ∂ is additive, $\partial(C_1 + C_2) = \partial(C_1) + \partial(C_2)$, and so defines a homomorphism from $C_2(\mathscr{K})$ to $C_1(\mathscr{K})$ and a homomorphism from $C_1(\mathscr{K})$ to $C_0(\mathscr{K})$.*

For example, consider the 1-chain $a + b$ on the complex $\mathscr{K}$ of Figure 23.2. Using the incidence coefficients given in table (a) of Figure 23.3, $\partial(a + b) = P + U$. Note that the vertex T is not in $\partial(a + b)$ because it has a total incidence with $(a + b)$ of 2, which is even. Both P and U have a total incidence of 1 with $(a + b)$, while the other vertexes have incidence 0. Note also that $\partial(a) = T + U$, $\partial(b) = P + T$, $\partial(a) + \partial(b) = T + U + P + T = P + U = \partial(a + b)$. This is an example of the additivity of the boundary operator. For another example, consider the 2-chain $(A + D)$. We find from table (b) of Figure 23.3 that $\partial(A + D) = (b + c + d + e + g + l)$.

Turning to the complex $\mathscr{L}$ of Figure 23.4 we find something interesting. The boundary of the simplex a is empty, $\partial(a) = \varnothing$, since the only vertex P has total incidence 2 with a. Similarly, $\partial(A) = \varnothing$, since both edges a and b are incident twice with the face A.

The proof of additivity for the boundary operator is the same as the corresponding proof in §15, with the sole difference that instead of the number of times a given $(k-1)$-simplex S is contained in a given k-chain C, one speaks of the *total incidence* of S in C, the total incidence being the sum of the incidence coefficients of S with the simplexes of C.

Definition

*A **k-cycle** ($k = 1, 2$) is a k-chain with null boundary. By convention, all 0-chains are called 0-cycles. A **k-boundary** ($k = 0, 1$) is a k-chain that is the boundary of a $(k+1)$-chain. By convention, only the null 2-chain is considered a 2-boundary. The k-cycles, $Z_k(\mathcal{K})$, and the k-boundaries, $B_k(\mathcal{K})$, are both groups in themselves.*

On account of their geometric significance, these two types of chains are of special importance. The homology groups involve only cycles and boundaries.

Theorem

Every boundary is a cycle.

This is obvious for 0-boundaries and 2-boundaries. For 1-boundaries it follows from the fact that for all 2-chains C, $\partial(\partial(C)) = \varnothing$. To prove this identity it suffices, by additivity, to confirm it for all 2-simplexes, that is, when C is a polygon, since every 2-chain is a sum of 2-simplexes. When C is a polygon, $\partial(\partial(C))$ consists of those vertexes of C that are incident on an odd number of edges of C. Since every vertex of a polygon has total incidence two on the edges of the polygon, $\partial(\partial(C)) = \varnothing$.

Exercises

7. Verify these relationships for the complex $\mathcal{K}$ of Figure 23.2. In each case draw the assembled cube, label the vertexes, and mark the chains involved.

(a) $\partial(B + D + E + F) = a + b + d + e + f + h + j$

(b) $\partial(b + c + k + f + l + g) = \varnothing$

(c) $\partial(d + c + i + h + j) = \partial(a) = T + U$

8. For the complex $\mathcal{K}$ of Figure 23.2, find

(a) a 2-chain that is not a 2-cycle

(b) a 1-chain that is not a 1-cycle

(c) a 2-cycle that is not a 2-boundary

(d) a 1-boundary that is not $\varnothing$

For the complex $\mathcal{L}$ of Figure 23.4, show that

(e) every 2-chain is a 2-cycle

(f) every 1-chain is a 1-cycle

(g) every 1-boundary is $\varnothing$

Somewhat later (§27) we will prove that every 1-cycle on $\mathcal{K}$ is a 1-boundary. In contrast, on $\mathcal{L}$ find

(h) a 1-cycle that is not a 1-boundary

(i) Does $\mathcal{L}$ have a 2-cycle that is not a 2-boundary?

9. Prove that in every connected complex $\mathcal{K}$, a 0-chain is a 0-boundary if and only if it has an even number of vertexes.

10. Prove that the k-chains of a complex form a group, in fact an idemgroup. Prove that the boundary operator is additive. Prove that the k-boundaries and k-cycles also form idemgroups.

Definition

Two k-chains ($k = 0, 1, 2$), C_1 and C_2, are called **homologous**, written $C_1 \sim C_2$, when $C_1 + C_2$ is a k-boundary. The relation of homology has the following properties:

(a) $C \sim C$ for every k-chain.

(b) If $C_1 \sim C_2$, then $C_2 \sim C_1$.

(c) If $C_1 \sim C_2$ and $C_2 \sim C_3$, then $C_1 \sim C_3$.

(d) If $C_1 \sim C_2$ and $C_3 \sim C_4$, then $C_1 + C_3 \sim C_2 + C_4$.

The group $Z_k(\mathcal{K})$ of k-cycles, when homology is used in place of equality, forms a group $H_k(\mathcal{K})$ called the **kth homology group** of $\mathcal{K}$.

Examples

1. *The cube.* We will compute the homology groups of the complex $\mathcal{K}$ described in Figure 23.2.

$H_0(\mathcal{K})$. All 0-chains are cycles. Since $\mathcal{K}$ is connected, by Exercise 9, exactly half of these are boundaries, the ones with an even number of vertexes. The 0-chains thus divide into two sets of mutually homologous cycles—all those with an even number of vertexes are homologous to each other, and all those with an odd number of vertexes are homologous. In effect, in the homology group $H_0(\mathcal{K})$ where homology takes the place of equality, there are just two elements, one representing all the cycles with an even number of vertexes, one representing all the cycles with an odd number of vertexes. Let these be represented by the symbols ε and θ (for even and odd); then the addition table for $H_0(\mathcal{K})$ is as follows:

	ε	θ
ε	ε	θ
θ	θ	ε

In summary: even plus even and odd plus odd are both even, while even plus odd is odd. Note that $H_0(\mathcal{K})$ is isomorphic to $\mathscr{C}_2$. Characteristically, the homology group is much smaller and simpler than the chain group.

$H_1(\mathcal{K})$. It is a fundamental fact about the sphere that in every complex equivalent to the sphere (including the cube $\mathcal{K}$) every 1-cycle is a 1-boundary. (See the fundamental lemma of §16.) For the cube this can be proven by the tedious process of inspecting all 64 1-cycles. However, no sane person would attack the problem this way. In the next section we find a simple means for computing all the homology groups of all the surfaces, from which this fact will follow as a trivial consequence. Let us assume, until this proof is supplied, that every 1-cycle on $\mathcal{K}$ is a 1-boundary. Then all 1-cycles are homologous to the null cycle $\varnothing$. The homology group $H_1(\mathcal{K})$ consists of a single element and so is isomorphic to the trivial group $\mathscr{C}_1$.

$H_2(\mathcal{K})$. The two-dimensional homology group is easily computed. There is only one 2-boundary by convention ($\varnothing$) and besides $\varnothing$ only one 2-cycle, namely the sum $\mu = A + B + C + D + E + F$. (Why?) Since μ is not homologous to $\varnothing$, $H_2(\mathcal{K})$ consists of just these two elements. You will easily verify that $H_2(\mathcal{K}) \cong \mathscr{C}_2$.

2. *The torus.* Among all the surfaces the sphere, and therefore the cube $\mathcal{K}$, has the simplest homology. To see something a little more interesting, turn to the complex $\mathscr{L}$ of Figure 23.4. The homology of this complex is quite easy to compute due to the circumstance that every k-chain ($k = 1, 2$) has boundary $\varnothing$. From this it follows first that every chain is a cycle, so that the homology groups include all the chains, and second that the only boundary is $\varnothing$. Therefore the relation of homology is the same as equality in this case.

Thus the homology groups of $\mathscr{L}$ are the *same* as the chain groups. We see that $H_0(\mathscr{L})$ contains just two elements $\emptyset$ and P and hence is isomorphic to $\mathscr{C}_2$. Likewise, $H_2(\mathscr{L})$ consists of just the two elements $\emptyset$ and A and so is also isomorphic to $\mathscr{C}_2$. These two groups are the same as with the cube. A difference appears in the first homology group $H_1(\mathscr{L})$, which contains four elements $\emptyset$, a, b, and $c = a + b$. A glance at the addition table given at the beginning of this section leads immediately to the interesting conclusion that the first homology group of the torus is isomorphic to Klein's four group!

Exercises

11. Prove that the relationship of homology has all the algebraic properties ascribed to it in the above definition, and that the k-cycles still form a group when equality is replaced by homology. Show that the homology groups are idemgroups.

12. Compute the homology groups of the following complexes.

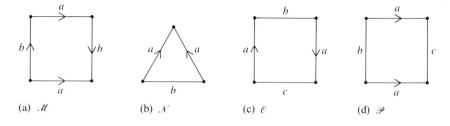

(a) $\mathscr{M}$ (b) $\mathscr{N}$ (c) $\mathscr{O}$ (d) $\mathscr{P}$

13. Since a graph has no 2-simplexes, it has only zero- and one-dimensional homology groups. Compute the homology groups of the following graphs.

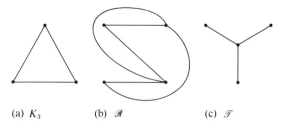

(a) K_3 (b) $\mathscr{R}$ (c) $\mathscr{T}$

§25 INVARIANCE

The homology groups of a topological space are among the most complicated constructions in all mathematics. To appreciate this, let us review the steps of the construction. (1) Given a topological space, we must first divide the space into polygons, edges, and vertexes identified in some way—in a word, we must view the topological space as a complex. (This is the geometric step of the construction.) (2) Given a complex, we extract the incidence coefficients with which we define the chain groups and the boundary operator. (This is the combinatorial step of the construction.) (3) Finally we isolate the subgroups of cycles and boundaries and define the homology groups. (This is the algebraic step of the construction.)

The main question is, Is it all worth it? Can the homology groups really supply *useful* information about topological spaces? This question will be answered gradually (in the affirmative) by the examples and applications in the rest of the book. In this section we want to answer an even more fundamental question: How can the homology groups tell anything *at all* about the topological space? Isn't the construction so long and complicated, so filled with the choices of simplexes and complexes, that in the end the groups have nothing to do with the original space? Although taking an extreme position, the question posed this way is important. The construction of the homology groups does depend on making a choice. In Step 1 one must choose a complex equivalent to the given topological space. Once that complex is chosen, however, the remaining steps of the construction 2 and 3 follow mechanically. Nonetheless it seems plausible that a given topological space, having different complexes, might also have different sets of homology groups. The sphere, for example, has triangulations and gratings galore, not to speak of other polyhedra or its normal form. Since these different complexes have wildly different sets of simplexes and incidence coefficients, it seems only reasonable that they should lead to different homology groups. And if the homology groups are different, what can they possibly tell us about the original unchanging topological space?

The solution to this problem is provided by the invariance theorem. With some restrictions, this theorem states that *no matter how the topological space is cut up to make a complex, the homology groups come out the same.* Thus a given topological space has only *one* set of homology groups, and *these groups are independent of the manner in which the topological space is viewed as a complex.* It follows that, whatever information the homology groups contain, this information is of a purely topological nature unaffected by the geometric, combinatorial, and algebraic steps of the construction. For this reason the invariance theorem is truly the fundamental theorem of homology. Without the invariance of the homology groups under the choice of complex, the homology groups would be filled with information about the complex irrelevant to the topological space. *With invariance,* we can be confident that the information carried by the homology groups concerns the topological space alone, even though we have no idea yet exactly what that information is!

In this book we will not prove the invariance theorem in complete generality. Instead we prove an invariance theorem for triangulations of surfaces that is broad enough to cover our applications.

Invariance Theorem

The homology groups associated with a triangulation $\mathcal{K}$ of a compact, connected surface $\mathcal{S}$ are independent of $\mathcal{K}$. In other words, the groups $H_0(\mathcal{K})$, $H_1(\mathcal{K})$, and $H_2(\mathcal{K})$ do not depend on the simplexes, incidence coefficients, or anything else arising from the choice of the particular triangulation $\mathcal{K}$; they depend only on the surface $\mathcal{S}$ itself.

The idea of the proof is to show that the homology groups of the triangulation $\mathcal{K}$ are the same as the homology groups of the *plane model $\mathcal{N}$* of the surface $\mathcal{S}$. To prove this, it will suffice to follow the proof of the classification theorem (§21) through all its six steps, showing that each step does not change the homology groups. Since at the beginning of the proof we have the triangulation $\mathcal{K}$, while at the end $\mathcal{K}$ has been transformed to the plane model $\mathcal{N}$, it will follow that they share the same homology groups, proving the theorem. The execution of this idea depends on the following lemma.

Lemma

Let $\mathcal{K}$ be any complex. Let $\mathcal{K}^+$ be the complex obtained from $\mathcal{K}$ by drawing a single new edge dividing a single polygon of $\mathcal{K}$ into two polygons. Then the homology groups of $\mathcal{K}$ and $\mathcal{K}^+$ are the same.

Figure 25.1 gives an example. Note that $\mathcal{K}^+$ not only has two new 2-simplexes (B and C) where $\mathcal{K}$ had one, and a new edge g separating B from C, but also $\mathcal{K}^+$ may have two new vertexes (P and Q), the vertexes of g, each of which divides an edge of $\mathcal{K}$ in two. The proof divides into a number of cases, according to how many new vertexes $\mathcal{K}^+$ has. Since the proofs in all cases are similar, we will be content to give the proof for the case described in Figure 25.1.

Let us begin by noting that certain combinations of the new elements of $\mathcal{K}^+$ correspond to the old elements in $\mathcal{K}$. Thus $B + C$ corresponds to A,

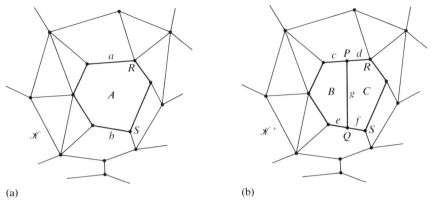

Figure 25.1

$c + d$ corresponds to a, and $e + f$ corresponds to b. This correspondence establishes an isomorphism from the chain groups of $\mathcal{K}$ into the chain groups of $\mathcal{K}^+$. It is obvious in this correspondence that two homologous chains in $\mathcal{K}$ always correspond to homologous chains on $\mathcal{K}^+$. Furthermore we will show that every cycle on $\mathcal{K}^+$ is homologous to a cycle that contains the new elements of $\mathcal{K}^+$ only in those combinations that correspond to elements from $\mathcal{K}$. Thus when equality is replaced by homology, the correspondence established between the chain groups becomes an isomorphism between the homology groups, proving the lemma.

Considering zero-dimensional homology first, observe that on $\mathcal{K}^+$, $P \sim R$ and $Q \sim S$, since $\partial(d) = P + R$ and $\partial(f) = Q + S$. Therefore every 0-cycle containing P or Q is homologous to a cycle that does not contain P or Q, that is, a cycle corresponding to a 0-cycle on $\mathcal{K}$. This concludes the proof for zero-dimensional homology.

Turning to one-dimensional homology, let λ be a 1-cycle on $\mathcal{K}^+$. There are three ways that λ might not correspond to a cycle on $\mathcal{K}$: (1) λ might contain c or d but *not* both, (2) λ might contain e or f but *not* both, and (3) λ might contain g. Suppose λ satisfies (1). Then because λ is a cycle, λ must also contain g (otherwise $\partial(\lambda)$ would contain P). Then also λ must contain e or f but not both (otherwise $\partial(\lambda)$ would contain Q). Thus λ actually satisfies (1), (2), and (3). Similar arguments beginning with the assumption of (2) or (3) lead us to conclude that either λ satisfies none of the conditions (1), (2), or (3) and therefore corresponds to a cycle on $\mathcal{K}$, or else λ satisfies all three conditions. In the latter case, consider $\mu = \lambda + \partial(B)$. The chain μ is a 1-cycle $\partial(\mu) = \partial(\lambda) + \partial(\partial(B)) = \varnothing + \varnothing$, homologous to λ ($\mu + \lambda = \partial(B)$) and μ satisfies none of the conditions (1), (2), or (3). Thus λ is homologous to a 1-cycle that corresponds to a cycle on $\mathcal{K}$. This completes the proof for one-dimensional homology.

Finally for dimension two, observe that a 2-cycle T on $\mathscr{K}^+$ contains both B and C or neither (otherwise $\partial(T)$ contains g). Thus T must correspond to a 2-cycle on $\mathscr{K}$. This completes the proof of the lemma.

This lemma may be regarded in two ways. We may consider the complex $\mathscr{K}$ as given, the complex $\mathscr{K}^+$ being obtained from it by subdividing one polygon of $\mathscr{K}$. Or we may consider the complex $\mathscr{K}^+$ as given, the complex $\mathscr{K}$ then being obtained from $\mathscr{K}^+$ by sewing together two polygons of $\mathscr{K}^+$ to make a single polygon for $\mathscr{K}$. In either case the lemma asserts that $\mathscr{K}$ and $\mathscr{K}^+$ have the same homology groups. In other words, we conclude from the lemma that neither the operation of cutting a polygon in two nor the operation of sewing two polygons together along an edge changes the homology of a complex. Now the proof of the classification theorem of §21 is nothing more than a long series of these cutting and pasting operations. For example, Step 1 of the proof involves a series of pasting operations leading to a model for the surface consisting of a single polygon. Our lemma applies to each of these pasting operations. Subsequent steps involve both cutting and pasting, while the model is reduced to one of the standard forms. In each case, with one exception, our lemma implies that the homology of the resulting complex is unchanged. The exception is the pasting operation used in Step 2, but here a separate argument (Exercise 1) leads to the same conclusion: the homology of the complex is not changed. It follows that no matter what triangulation is used, the homology groups of a surface are the same as those of its plane model. Therefore the homology groups are independent of the triangulation.

As it turns out, we have proved much more than the invariance theorem itself. In the first place, we have the following principle.

Invariance Principle

The homology groups of a complex are not altered by either cutting a polygon in two or pasting together two polygons along an edge.

This invariance principle is an important step in the proof of a general invariance theorem. Early investigations of invariance depended on this principle and the notion of a common refinement of two triangulations. Given two triangulations of a topological space, a common refinement is any third triangulation that can be obtained from both of the given triangulations by cutting operations alone. In the last of his pioneering papers on combinatorial topology (1904), Poincaré made the following famous conjecture.

Hauptvermutung

Any two triangulations of a topological space have a common refinement.

The Hauptvermutung (principle conjecture) was stated by Poincaré only for the analogue of surfaces in three dimensions: three-dimensional manifolds. It was proven much later by Moise (1948). Still later it was verified for *all* triangulable spaces of dimension two by Papakyriakopolous (1963). The Hauptvermutung and the invariance principle together imply the following theorem.

General Invariance Theorem

Let $\mathcal{T}$ be a triangulable space of dimension two, that is, a complex composed of simplexes of dimension two or less. Then the homology groups of $\mathcal{T}$ are independent of the choice of triangulation.

Unfortunately the Hauptvermutung is false for general topological spaces of dimension greater than two. Therefore a general invariance theorem for higher dimensions cannot be proven this way. A counterexample was found by Milnor (1961) for dimensions six and greater. It is still unknown whether the Hauptvermutung is true for *manifolds* of dimension greater than three. The first proof of invariance by Alexander (1926) used a different technique, simplicial approximation, which we shall discuss in Chapter Six. The short proof of invariance for surfaces presented here is possible only by virtue of the classification theorem. Here is another important consequence of this proof.

Theorem

The homology groups of a compact, connected surface are the homology groups of its normal form.

Using this theorem it is easy to compute the homology groups of all the compact, connected surfaces. For example, consider the sphere (Figure 19.2). There are two vertexes in the model, but they are homologous, and therefore $H_0 \cong \mathscr{C}_2$. There is one edge, but there are no 1-cycles except $\varnothing$, and therefore $H_1 \cong \mathscr{C}_1$. The one face is a 2-cycle, and therefore $H_2 \cong \mathscr{C}_2$. These results agree with those obtained from the cube $\mathscr{K}$ in the previous section.

Consider for a moment the significance of the fact that H_1 for the sphere is the trivial group. This means that every 1-cycle is homologous to $\varnothing$, or that every 1-cycle is a boundary. This conclusion holds not only for the normal form, but, according to the invariance theorem, for all triangulations of the sphere; more generally, according to the argument of this section, the conclusion holds for any complex equivalent to the sphere that can be reduced to the normal form by cutting and pasting. These include, among the many complexes equivalent to the sphere, the gratings used in Chapter Three. Thus this result is a vast generalization of the fundamental lemma of §16.

Exercises

1. Complete the proof of the invariance theorem by proving that the homology groups are not affected by the type of pasting operation used in Step 2 of the proof of the classification theorem.

2. The normal forms of all the compact, connected surfaces consist of a single polygon whose edges are identified in pairs and whose vertexes are all identified together. What does this imply concerning the groups H_0 and H_2 of all these spaces?

3. Compute the homology of the projective plane, and compare it with the homology of the sphere. Compare also the homology of the torus with the homology of the Klein bottle.

4. Verify that the invariance theorem holds for surfaces with boundary. Compute the homology of the disk, cylinder, and Möbius strip. Compare these results with the homology of the corresponding surfaces without boundary.

5. Compute the homology of the following two types of space for low values of n ($n = 1, 2, 3, 4$).

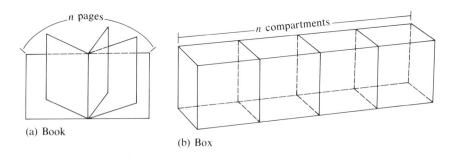

(a) Book

(b) Box

§26 BETTI NUMBERS AND THE EULER CHARACTERISTIC

According to the invariance theorem, algebraic properties of the homology groups must reflect topological properties of the original topological space. As yet, however, we have no idea what these properties on either side might be. In order to unlock the secrets of the homology groups, it will be necessary to learn more about groups, especially idemgroups. That is what we propose to do in this section. We begin with the notion of dependence.

Definition

*Let G be an idemgroup. (In this section all groups will be idemgroups.) Let $A = \{y_1, y_2, \ldots, y_n\}$ be a subset of G. An element x of G **depends** on A if x is equal to a sum of elements from A, that is,*

$$x = a_1 y_1 + a_2 y_2 + \cdots + a_n y_n$$

where each coefficient $a_1, a_2, \ldots, a_n$ is either zero or one.

The expression for x in the above equation is called a **linear combination** of the elements of A. Another way to phrase the definition of linear dependence is to say that x depends on A if x is equal to a linear combination of elements from A. In this connection it is perfectly possible to consider values for the coefficients $a_1, a_2, \ldots, a_n$ larger than zero, or one. In idemgroups, however, larger coefficients are useless, since $2y = y + y = 0, 3y = y + 2y = y$, $4y = 0$, and so forth.

The theory of dependence is based on the following fundamental properties:

(a) Every element in the set A depends on A.

(b) If every element in the set A depends on a second set B, then any element x depending on A also depends on B.

(c) If x depends on the set $\{y_1, y_2, \ldots, y_n\}$ but does not depend on the set $\{y_1, y_2, \ldots, y_{n-1}\}$, then 0_n depends on the set $\{y_1, y_2, \ldots, y_{n-1}, x\}$.

Proofs of the first two properties are easy to supply (Exercise 1). The third property is called the **exchange axiom** and is proved as follows. Let x be dependent on $\{y_1, y_2, \ldots, y_n\}$. Then x is a linear combination of the elements $y_1, y_2, \ldots, y_n$, so that $x = a_1 y_1 + a_2 y_2 + \cdots + a_n y_n$, where each coefficient $a_1, a_2, \ldots, a_n$ is a zero or a one. If the coefficient a_n is zero, then x is also dependent on the smaller set $\{y_1, y_2, \ldots, y_{n-1}\}$. Supposing that x is *not* dependent on $\{y_1, y_2, \ldots, y_{n-1}\}$, it follows that a_n equals one. Thus

$$x = a_1 y_1 + a_2 y_2 + \cdots + a_{n-1} y_{n-1} + y_n$$

Adding x and y_n to both sides of this equation we get

$$y_n = a_1 y_1 + a_2 y_2 + \cdots + a_{n-1} y_{n-1} + x$$

Therefore y_n depends on the set $\{y_1, y_2, \ldots, y_{n-1}, x\}$. This proves the exchange axiom.

Definition

Let G be an idemgroup and A a subset of G. The subset A **spans** *G if every element of G depends on A. The subset A is* **independent** *if no element of A depends on the other elements of A. An independent set that spans G is called a* **basis** *for G.*

Theorem 1

Let A be a finite set spanning G. Then A contains a subset B that is a basis for G.

Let $A = \{y_1, y_2, \ldots, y_n\}$. Suppose that A is not yet a basis, meaning that A is not independent. Then some element of A, let us say y_1, depends on the set $A' = \{y_2, y_3, \ldots, y_n\}$ consisting of the remaining elements of A. Now every element of A depends on the subset A' (using property (a)). Therefore by property (b), every element that depends on A also depends on A'; thus A' still spans G. Supposing that A' is not independent, we can throw out some element of A' that depends on the others, obtaining a subset A''. By the argument just given, A'' still spans G. Continuing in this way we eventually obtain an independent subset B of A that spans G. B is the desired basis.

Observe that in the proof of Theorem 1 the basis B is constructed by throwing out of A the elements that are not needed to span G, the elements that depend on other elements in the set. The basis B is thus a minimal subset of A still spanning G.

Theorem 2

Any two bases for G have the same number of elements.

Let $A = \{x_1, x_2, \ldots, x_m\}$ and $B = \{y_1, y_2, \ldots, y_n\}$ be two bases for G. We will prove that $m = n$. The element x_1 is dependent upon B, as are all elements of G. Choose the element y_k of B so that x_1 depends on the subset $\{y_1, y_2, \ldots, y_k\}$ but does not depend on $\{y_1, y_2, \ldots, y_{k-1}\}$. Such an element y_k must always exist; k is the smallest number such that x_1 is dependent on $\{y_1, y_2, \ldots, y_k\}$. By the exchange axiom, y_k is dependent on $\{y_1, y_2, \ldots, y_{k-1}, x_1\}$. Let B_1 be the set B with x_1 in place of y_k, that is, $B_1 = \{x_1, y_1, y_2, \ldots, y_{k-1}, y_{k+1}, \ldots, y_n\}$. Every element of B depends on the set B_1; hence by property (b) B_1 spans G. Thus we have obtained a spanning set B_1 that differs from B by the exchange of one element from A.

Consider next the element x_2 from A. Since B_1 spans G, x_2 depends on B_1. Let $\{x_1, y_1, y_2, \ldots, y_i\}$ be an initial portion of B_1 so that x_1 depends on this set but not on this set minus y_i. Such a set must include an element y, since the set A (of x's) is independent. Then just as in the preceeding paragraph, the set B_2, obtained by replacing y_i in B_1 by x_2, still spans G. Continuing in this way we obtain a sequence $B_1, B_2, \ldots, B_m$ of sets all spanning G, each B_j being obtained from its predecessor by the exchange of an element from the original set B with the element x_j from A. This replacement is made possible each time by the fact that x_j must depend on the preceeding set B_{j-1} but can't depend only on the elements from A already exchanged into B_{j-1} because A is independent. Eventually we reach the set B_m, which must contain A as a subset. Thus $m \leq n$. Note that so far the proof has only made use of the fact that B spans G while A is independent. Reversing the roles of A and B, using the fact that A spans G while B is independent, we find that also $n \leq m$. Therefore $m = n$ as desired.

These two theorems fit together in the following way. Suppose that G is a finite idemgroup, for example a chain group or homology group of a finite complex associated with a compact topological space. By Theorem 1, G contains bases, while according to Theorem 2, all these bases contain the same number of elements. The number of elements in a basis of G is called the **rank** of G. The rank is a crude measure of the size of the group G. We shall find that the ranks of the homology groups have important geometric interpretations.

Exercises

1. Prove properties (a) and (b) of dependence. Prove that the zero of G is dependent on every set A.
2. Prove that a set A is independent if and only if whenever a linear

combination $a_1 x_1 + a_2 x_2 + \cdots + a_n x_n$ of elements from A equals zero, then all the coefficients $a_1, a_2, \ldots, a_n$ are zero.

3. Let $B = \{y_1, y_2, \ldots, y_n\}$ be a basis for G. Prove that an element x of G is equal to a unique linear combination of elements from B. Conclude that if the rank of G is n, then G has 2^n elements.

4. Let H be a subgroup of G. Prove that the rank of H is less than or equal to the rank of G. Prove that if the rank of H equals the rank of G, then $H = G$.

5. Prove that if the rank of G is n, then any independent set of n elements is a basis.

6. Let A be an independent subset of G. Show that there exists a basis for G containing A.

Let us now consider the ranks of the chain groups and homology groups associated with a complex. Let $\mathscr{K}$ be a finite complex. Considering chain groups first, it is obvious that the set of k-simplexes is a basis for the group $C_k(\mathscr{K})$, since every chain is a sum of simplexes and the set of simplexes is independent. Thus the rank c_k of $C_k(\mathscr{K})$ is simply the number of k-simplexes. This number, although of combinatorial interest, is clearly not of topological interest. Passing to the homology groups themselves, the rank h_k of $H_k(\mathscr{K})$ must, by virtue of the invariance theorem, have an intrinsic topological significance, at least for surfaces. The numbers h_k are called the **Betti numbers** of the complex.

The zeroth Betti number h_0 equals 1 for all connected surfaces since, in a complex on a connected surface, any two vertexes are always homologous. More generally, an arbitrary surface is always the union of a number of connected pieces called **components**. The zeroth Betti number is the number of these components (Exercise 7). The first Betti number h_1 is called the **connectivity number** of the surface. It gives the largest number of closed curves that can be drawn on the surface *without* dividing the surface into two or more pieces. (This number is zero for the sphere.) The proof of this is beyond the scope of this book; it is a vast generalization of the Jordan curve theorem. The second Betti number h_2 is one for all compact, connected surfaces (Exercise 10) and so appears to carry no topological information. However, this is a peculiarity of the homology considered in this section and will change with the introduction of integral homology in §29.

The Betti numbers were first defined by Betti (1870) directly in terms of the topological interpretation given them in the preceeding paragraph. Their later appearance in the combinatorial work of Poincaré increased their usefulness immensely and accounts for their importance in topology. Poincaré himself did not define the homology groups but used cycles and boundaries to define the Betti numbers directly. Most of the early applications of homology were based directly on these numbers. The introduction of the language of group theory occurred much later (1925–1935) under the influence of the great algebraist Emmy Noether.

Exercises

7. Prove that, for compact surfaces, the zeroth Betti number is the number of components of the surface, where a component is a connected subset of the surface, such that any larger containing subset is not connected.

8. Show that the first Betti number is $2k$ for the connected sum of k tori and k for the connected sum of k projective planes. In each case show that it is possible to draw h_1 curves on the surface without disconnecting the surface.

9. Show that the first Betti number of a surface with boundary equals the first Betti number of the corresponding surface without boundary plus the number of boundary curves minus one.

10. Determine the second Betti number for all compact, connected surfaces with and without boundary.

11. Determine the Betti numbers of the n-page book and the box with n compartments (see Exercise 5 of §25).

12. Let $\mathcal{G}$ be a connected graph. Show that $h_0 = 1$ for $\mathcal{G}$. Let V and E be the numbers of vertexes and edges of $\mathcal{G}$. Let $\mathcal{T}$ be a spanning tree in $\mathcal{G}$. For each edge e not part of $\mathcal{T}$, show that there is a unique cycle on $\mathcal{G}$ made up of e plus some edges from $\mathcal{T}$. Show that the set of cycles thus obtained is independent and spans the space of 1-cycles on $\mathcal{G}$. Conclude that $h_1 = E - V + 1$.

13. Let $\mathcal{G}$ be a graph. Prove that h_0 is the number of components of $\mathcal{G}$, while $h_1 = E - V + h_0$.

14. Find the Betti numbers of the following graphs. Find a basis of the group of 1-cycles explicitly using a spanning tree.

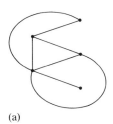
(a)

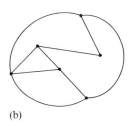
(b)

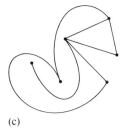
(c)

THE EULER CHARACTERISTIC

For any complex equivalent to a sphere, the number of vertexes minus the number of edges plus the number of polygons is always two. On this account the number two is called the **Euler characteristic** of the sphere. In many ways this property is *the* critical topological property of the sphere, and the number two appears time after time in applications of topology to the sphere. To understand this we shall establish similar results for the other compact, connected surfaces. The proof requires further study of the concept of the rank of a group. This should not be surprising, since the numbers of vertexes, edges, and polygons in a complex are the ranks of the associated chain groups.

Definition

Let G and H be groups, and f a homomorphism from G to H. The **kernel** *of f, $K(f)$, is the set of elements of G such that $f(x)$ is zero in H. The* **range** *of f, $R(f)$, is the set of elements $f(x)$ in H (letting x range over all the elements of G). Both $K(f)$ and $R(f)$ are subgroups, the former of G, the latter of H (Exercise 15). The rank k of $K(f)$ is called the* **nullity** *of f; the rank r of $R(f)$ is called the* **rank** *of f.*

As an example, consider the boundary operator ∂, which is, after all, a group homomorphism from the group of k-chains C_k on some complex $\mathscr{K}$ to the group of $(k-1)$-chains, C_{k-1} ($k = 1, 2$). The kernel of ∂ is the group of k-cycles, while the range of ∂ is the group of $(k-1)$-boundaries. The following theorem gives a very important relationship between the rank and nullity.

Theorem 3

Let f be a homomorphism from the group G to the group H. Then the rank of f plus the nullity of f equals the rank of G.

Intuitively it is clear that the larger the kernel of f (i.e., the more elements f transforms to the zero of H), the smaller the range of f (i.e., the fewer elements f can send to nonzero elements of H). Theorem 3 simply makes this numerically precise, showing how the rank of f must decrease if the nullity increases, since the two sum to the rank of G.

For the proof let $\{x_1, x_2, \ldots, x_k\}$ be a basis for $K(f)$, k being the nullity of f. Let $\{x_{k+1}, x_{k+2}, \ldots, x_n\}$ be elements of G that when added to $\{x_1, x_2, \ldots, x_k\}$ produce a basis for G (Exercise 6), n being the rank of G. We must show that $n - k$ equals r, the rank of f. This is accomplished by showing that the $n - k$ elements of H, $y_k = f(x_k)$, $y_{k+1} = f(x_{k+1})$, ..., $y_n = f(x_n)$, form a basis for $R(f)$. Thus we shall prove (a) that every element of $R(f)$ depends on $\{y_{k+1}, \ldots, y_n\}$ and (b) that the set $\{y_{k+1}, \ldots, y_n\}$ is independent.

To prove (a), let y be any element of $R(f)$. Then y equals $f(x)$ for some element x in G. As an element of G, x can be expressed as a linear combination $a_1 x_1 + a_2 x_2 + \cdots + a_n x_n$ of the basis $\{x_1, x_2, \ldots, x_n\}$ of G. Then

$$y = f(x) = f(a_1 x_1 + \cdots + a_n x_n) = a_1 f(x_1) + a_2 f(x_2) + \cdots + a_n f(x_n)$$
$$= a_{k+1} y_{k+1} + \cdots + a_n y_n$$

since $f(x_1), f(x_2), \ldots, f(x_k)$ are all zero. This proves that y depends on $\{y_{k+1}, \ldots, y_n\}$.

To prove (b), suppose that the set $\{y_{k+1}, \ldots, y_n\}$ is not independent. Then, according to Exercise 2, there is a linear combination $a_{k+1} y_{k+1} + \cdots + a_n y_n$ equal to zero in which at least one coefficient is *not* zero. Then

$$f(a_{k+1} x_{k+1} + \cdots + a_n x_n) = a_{k+1} f(x_{k+1}) + \cdots + a_n f(x_n)$$
$$= a_{k+1} y_{k+1} + \cdots + a_n y_n = 0$$

In other words, the element $a_{k+1} x_{k+1} + \cdots + a_n x_n$ of G is in the kernel of f. Since $\{x_1, \ldots, x_k\}$ is a basis for $K(f)$, we have $a_{k+1} x_{k+1} + \cdots + a_n x_n = a_1 x_1 + \cdots + a_k x_k$ for some coefficients $a_1, a_2, \ldots, a_k$. Thus $a_1 x_1 + \cdots + a_k x_k + a_{k+1} x_{k+1} + \cdots + a_n x_n = 0$, where not all the coefficients in this expression equal zero. This contradicts the fact that $\{x_1, \ldots, x_k, x_{k+1}, \ldots, x_n\}$ is independent. Therefore the set $\{y_{k+1}, \ldots, y_n\}$ is independent as well. This completes the proof of the theorem.

As an example of the application of this theorem, let $\mathscr{K}$ be a complex. Let C_0, C_1, C_2 be the chain groups of $\mathscr{K}$, and let B_0, B_1, B_2 be the groups of boundaries and Z_0, Z_1, Z_2 the groups of cycles. The boundary operator provides two homomorphisms

$$C_2 \xrightarrow{\partial_1} C_1 \xrightarrow{\partial_2} C_0$$

for which $K(\partial_2) = Z_2$, $K(\partial_1) = Z_1$ and $R(\partial_2) = B_1$, $R(\partial_1) = B_0$. Recall that by convention, $Z_0 = C_0$ and $B_2 = \{\varnothing\}$ (the trivial group). Let c_0, c_1, c_2, b_0, b_1, b_2, and z_0, z_1, z_2 be the ranks of all these groups. Then according to Theorem 3,

$$c_2 = z_2 + b_1$$
$$c_1 = z_1 + b_0 \tag{1}$$

Definition

Let H be a subgroup of the group G. Two elements x and y of G are called **homologous (mod H)**, *written* $x \sim y$ *(mod H), if x equals y plus an element of H. The relation of homology (mod H) has all the algebraic properties of equality; therefore G remains a group when homology (mod H) is used in place of equality. The resulting group is called the* **quotient group** *of G by H and is written G/H.*

You will instantly recognize that this definition merely recapitulates in abstract form the construction of the homology groups. According to this definition, the homology groups are all quotient groups. To be precise, suppose $\mathscr{K}$ to be the same complex used in the paragraph above, and let H_0, H_1, and H_2 be the homology groups of $\mathscr{K}$. Then $H_0 = Z_0/B_0$, $H_1 = Z_1/B_1$ and $H_2 = Z_2/B_2$. The next theorem provides still more information on the ranks of all these groups

Theorem 4

Let H be a subgroup of G. Then the rank of G/H equals the rank of G minus the rank of H.

The proof is an almost exact imitation of the proof of Theorem 3. We first choose a basis $\{x_1, x_2, \ldots, x_k\}$ for H, where k is the rank of H, and then extend this as before to a basis $\{x_1, \ldots, x_k, x_{k+1}, \ldots, x_n\}$ for G, where n is the rank of G. We must prove that the rank of G/H is $n - k$. This will be accomplished by proving that $\{x_{k+1}, \ldots, x_n\}$ is a basis for G/H. Thus we must show (a) that every element of G depends (mod H) on $\{x_{k+1}, \ldots, x_n\}$ and (b) that $\{x_{k+1}, \ldots, x_n\}$ is independent (mod H). To prove (a), let x be any element of G. Then for suitable coefficients $a_1, \ldots, a_n$, $x = a_1 x_1 + \cdots + a_n x_n$, since $\{x_1, \ldots, x_n\}$ is a basis for G. Therefore, because $x_1, x_2, \ldots, x_k$ are all contained in H, we have $x \sim a_{k+1}x_{k+1}, + \cdots + a_n x_n$ (mod H). Thus x depends (mod H) on $\{x_{k+1}, \ldots, x_n\}$. To prove (b), suppose that $\{x_{k+1}, \ldots, x_n\}$ is *not* independent (mod H). Then there would be coefficients $a_{k+1}, \ldots, a_n$ not all zero such that $a_{k+1}x_{k+1} + \cdots + a_n x_n \sim 0$ (mod H). Since $\{x_1, \ldots, x_k\}$ is a basis for H, this implies that there exist coefficients $a_1, \ldots, a_k$ such that $a_{k+1}x_{k+1} + \cdots + a_n x_n = a_1 x_1 + \cdots + a_k x_k$. As in the proof of Theorem 3, this contradicts the independence of the basis $\{x_1, \ldots, x_n\}$ of G, so that $\{x_{k+1}, \ldots, x_n\}$ must be independent (mod H).

Applying Theorem 4 to the homology groups H_0, H_1, H_2 of the complex $\mathscr{K}$ we obtain the relations

$$h_0 = z_0 - b_0$$

$$h_1 = z_1 - b_1 \tag{2}$$
$$h_2 = z_2 - b_2$$

where h_0, h_1, and h_2 are the ranks of the homology groups. The equations (1) and (2) together with the consequences of the convention regarding Z_0 and B_2, namely

$$c_0 = z_0$$
$$b_2 = 0 \tag{3}$$

enable us to prove the following theorem.

Theorem (Poincaré)

Let the surface $\mathscr{S}$ be given as a complex $\mathscr{K}$. Let V, E, and F be the number of vertexes, edges, and faces in $\mathscr{K}$. Then the sum $V - E + F$ is a constant independent of the manner in which $\mathscr{S}$ is divided up to form the complex $\mathscr{K}$. This constant is called the **Euler characteristic** *of the surface and is denoted $\chi(\mathscr{S})$.*

The proof uses the fact that

$$c_2 = F$$
$$c_1 = E \tag{4}$$
$$c_0 = V$$

Thus

$$\begin{aligned}
V - E + F &= c_0 - c_1 + c_2 \\
&= z_0 - (z_1 + b_0) + (z_2 + b_1) \\
&= (z_0 - b_0) - (z_1 - b_1) + (z_2 - b_2) \\
&= h_0 - h_1 + h_2
\end{aligned}$$

According to the invariance theorem, the sum $h_0 - h_1 + h_2$ is independent of the complex $\mathscr{K}$, so this proves the theorem.

We can immediately compute the Euler characteristic of various surfaces using the knowledge we have already assembled concerning the Betti numbers. Thus the torus has characteristic zero, as does the Klein bottle. The projective plane has characteristic one, and the double torus has characteristic minus two. In general,

$$\chi(\mathscr{S}) = 2 - h_1$$

a result known as the **Euler-Poincaré formula**.

From the point of view of the sphere, an interesting fact emerges. All other compact, connected surfaces have characteristics smaller than two, since the sphere is the only compact, connected surface with zero connectivity number. Thus the Euler characteristic actually does characterize the sphere among compact, connected surfaces. To put it more dramatically, suppose you were imprisoned on a surface and your captors as part of their fiendish scheme of torture kept secret the nature of the imprisoning surface. Knowing topology, you would immediately divide the surface into polygons, edges, and vertexes and determine the Euler characteristic of the surface. If the result turned out to be two, you would have the relief of knowing that you were imprisoned on a sphere. Other characteristics might leave you in doubt. For example, a characteristic of zero would leave you unsure whether the prison was a torus or a Klein bottle. (Oh, the horror of learning, if the characteristic were one, that you were prisoner on a projective plane!)

The importance of the Euler characteristic for the sphere (and other surfaces as well) is not only that it characterizes, or helps characterize, the surface but also that it is easily computed, since it relates directly to the numbers of simplexes of any complex on the surface.

Exercises

15. Let f be a homomorphism from G to H. Prove that $K(f)$ and $R(f)$ are subgroups of G and H, respectively.

16. Determine the Euler characteristic of all compact, connected surfaces, with and without boundary. List all surfaces with and without boundary that have characteristics 2, 1, 0, and -1.

17. Determine the Euler characteristic of the n-page book, and the box with n compartments.

18. For a graph $\mathscr{G}$, the Euler characteristic is defined by $\chi(\mathscr{G}) = V - E$. Show by examples that $\chi(\mathscr{G})$ can have any integral value positive, negative, or zero.

19. Prove the Euler-Poincaré formula for graphs

$$V - E = h_0 - h_1$$

using Theorems 3 and 4. This gives a second proof of the result in Exercise 13.

20. Prove that among connected graphs, an Euler characteristic of one is characteristic of trees. Among graphs in general, those with $h_1 = 0$ are called **forests**. Explain the meaning of this term.

---

§27 MAP COLORING AND REGULAR COMPLEXES

As an application of the Euler characteristic, consider the following amusing problem first proposed over a hundred years ago. Given a map in the plane, the problem is to determine the minimum number of colors necessary to color it so that adjacent regions are colored with different colors. While there are some maps that can be colored with only two or three colors, many maps require four colors, including maps of just four countries (find one!). For many years it was an outstanding unsolved problem to prove that every map can be colored with just four colors. Many 'proofs' were discovered only to have flaws exposed later. The first such proof was published by Kempe in 1879. A flaw was uncovered by Heawood in 1890. By its very simplicity of statement, the problem attracted attention from amateur as well as professional mathematicians. Finally in 1976 the four color theorem was proved by Appel and Haken [1].

In the language of combinatorial topology, a map is simply a complex in which polygons are called countries. It makes sense to ask of *any* surface: What is the smallest number of colors needed to color faces of any complex equivalent to the surface, so that two faces sharing an edge are always colored with different colors? This is the aspect of the problem we will discuss here. It was first attacked by Heawood in 1890. Before presenting his results, it is important to point out the equivalence of the map coloring problem in the plane with the corresponding problem on the sphere. Every map in the plane can be placed on the sphere by adding a new country surrounding the given map (call this added country the "ocean") and then using stereographic projection. Conversely, every map on the sphere can be placed in the plane by cutting a disk out of one country and flattening the rest into the plane. It follows that if every planar map is four-colorable, then every map on the sphere is four-colorable, and vice versa.

Consider now any compact, connected surface $\mathscr{S}$. Let $\mathscr{K}$ be a complex on $\mathscr{S}$ with F faces, E edges, and V vertexes. The quantity $a = 2E/F$ is very important in map coloring. The numerator $2E$ can be regarded as the total number of edges counted face by face (because each edge is in two faces). Thus a is the *average number of edges per face*. The following lemma will explain the importance of a.

Lemma

Given the positive integer N, suppose $a < N$ for all complexes $\mathscr{K}$ on $\mathscr{S}$. Then N colors are sufficient to color all maps on $\mathscr{S}$.

The proof is by induction on the number of faces in $\mathcal{K}$. If there are fewer than N faces, that is, if $F < N$, then every face can be painted a different color. This takes care of the initial part of the induction proof. Now suppose that every complex with k faces is N-colorable, and let $\mathcal{K}$ be a complex with $F = k + 1$. We must prove that $\mathcal{K}$ is also N-colorable. Since the average number of edges per face is less than N, there is at least one face with fewer than N edges. Let $\mathcal{K}^-$ be a new complex formed by shrinking this face to a point, that is, by distributing this face among the adjoining faces as described in Figure 27.1. The complex $\mathcal{K}^-$ has k faces. By the induction assumption, $\mathcal{K}^-$ can be colored using N colors. The coloring of $\mathcal{K}^-$ leads to a coloring of $\mathcal{K}$, because the one face of $\mathcal{K}$ not already colored on $\mathcal{K}^-$ has *fewer than N adjoining faces* and so may be colored with one of the colors not present among the adjoining faces. This completes the proof of the lemma.

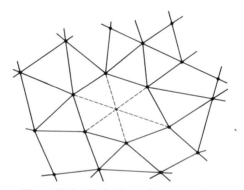

Figure 27.1   Shrinking a face to a point.

The argument of this lemma is fundamental to all coloring theorems. The problem with the conclusion from a topological point of view is that the quantity a is not related in any obvious way to the topological nature of the surface $\mathcal{S}$. Thus the lemma provides no link between the topological and the colorability properties of $\mathcal{S}$. However, since a involves both E and F, it makes sense to try and connect a with the Euler characteristic χ, which *is* an expression of the topological nature of $\mathcal{S}$. To do this, let $\mathcal{K}$ again be a complex on $\mathcal{S}$. We first eliminate from $\mathcal{K}$ all vertexes at which only two edges meet, by combining those two edges in one. This does not affect the colorability of the complex $\mathcal{K}$. Now every vertex lies on at least three edges of $\mathcal{K}$, so that $3V \le 2E$. From

$$\chi - F + E = V \le \frac{2E}{3}$$

we solve for E to obtain

$$E \leq 3(F - \chi)$$

Therefore

$$a = \frac{2E}{F} \leq 6\left(1 - \frac{\chi}{F}\right) \tag{1}$$

This gives an upper bound for a involving χ. We can apply it immediately to obtain the following theorem.

Theorem

Six colors are sufficient to color any map on the sphere or projective plane.

The sphere and projective plane are the two compact, connected surfaces whose characteristic is *positive*. For them (1) gives $a \leq 6(1 - \chi/F) < 6$. The theorem then follows immediately from the lemma.

This is a weak result for the sphere, since a four color theorem can be proved (although only with a great deal of difficulty). For the projective plane, on the other hand, this is exact. There exists a map on the projective plane (Figure 27.2) with six countries, each of which adjoins the other five. Therefore six colors are actually required to color every map on the projective plane, and we have solved the coloring problem for this surface.

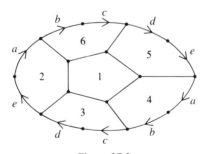

Figure 27.2

Historically the first surface for which the coloring problem was solved was the torus. Here (1) gives $a \leq 6 < 7$; therefore 7 colors are sufficient to color any map on the torus. (This argument applies equally well to the Klein bottle.) The map shown in Figure 27.3, discovered by Heawood, shows that 7 colors are required to color all maps on the torus.

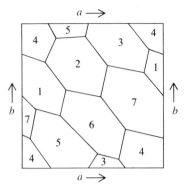

Figure 27.3

We consider next any surface $\mathscr{S}$ whose characteristic is less than zero. Let N stand for the number of colors we hope will color any map on $\mathscr{S}$. We need only consider maps with $F \geq N + 1$, for otherwise there are as many or more colors as countries. Then

$$\frac{1}{F} \leq \frac{1}{N + 1}$$

so that

$$\frac{\chi}{F} \geq \frac{\chi}{N + 1}$$

since $\chi \leq 0$. Now

$$a \leq 6\left(1 - \frac{\chi}{F}\right) \leq 6\left(1 - \frac{\chi}{N + 1}\right)$$

so that to ensure that $a < N$ we set

$$6\left(1 - \frac{\chi}{N + 1}\right) < N$$

If N satisfies this inequality, then, by the lemma, N colors will suffice to color all maps on $\mathscr{S}$. Solving this inequality we obtain

$$6N + 6 - 6\chi < N^2 + N$$

or

$$N^2 - 5N + (6\chi - 6) > 0$$

Therefore

$$N > \frac{5 + \sqrt{49 - 24\chi}}{2}$$

We are naturally interested in choosing the smallest N satisfying the inequality. The smallest N here is also the smallest N satisfying

$$N + 1 > \frac{7 + \sqrt{49 - 24\chi}}{2}$$

Now the *smallest integer* N satisfying this inequality is also the *largest integer* satisfying

$$N \leq \frac{7 + \sqrt{49 - 24\chi}}{2}$$

Thus we have proved the following theorem.

Theorem

Let $\mathscr{S}$ be any surface of characteristic $\chi \leq 0$. Then any map on $\mathscr{S}$ can be colored by N_χ colors,

$$N_\chi = \left[\frac{7 + \sqrt{49 - 24\chi}}{2} \right]$$

where the square brackets signify the largest integer less than or equal to the quantity inside.

This result was obtained by Heawood in 1890. The table below gives the values of N_χ for some values of χ.

χ	2	1	0	-1	-2	-3	-4	-5	-6	-7	-8	-9	-10
N_χ	4	6	7	7	8	9	9	10	10	10	11	11	12

It is important to realize that we have not proven that N_χ is the *minimum* number of colors sufficient to color all maps on these surfaces. Still it seems reasonable to make the following conjecture.

Heawood's Conjecture

N_χ is the minimum number of colors needed to color all maps on a surface of characteristic χ, where $\chi \leq 0$.

Note that N_χ gives the correct value for $\chi = 1$ and $\chi = 2$. The Heawood conjecture remained unverified for nearly eighty years. During this period progress was made. Heawood himself found the map in Figure 27.3, and others found maps settling the conjecture for other surfaces. In 1934 Franklin established the remarkable result that *only six colors are needed to color every map on the Klein bottle*, where the Heawood conjecture is *seven*. Finally in 1968, Ringel and Youngs, after a decade of work, were able to settle the conjecture in every case. Their long proof devolves into twelve cases according to the remainder when χ is divided by 12. The conjecture was confirmed in every case but that of the Klein bottle.

Exercises

1. How many colors are required in order to color all maps on a Möbius strip? Discuss the map coloring problem for other surfaces with boundary.

2. Let $\mathscr{K}$ be a triangulation of a compact, connected surface $\mathscr{S}$. Prove that

(a) $3F = 2E$

(b) $E = 3(V - \chi)$

(c) $V \geq \dfrac{7 + \sqrt{49 - 24\chi}}{2}$

3. Verify these lower bounds for V, E, and F for triangulations of the indicated surfaces.

sphere	$V \geq 4$	$E \geq 6$	$F \geq 4$
projective plane	$V \geq 6$	$E \geq 15$	$F \geq 10$
torus	$V \geq 7$	$E \geq 21$	$F \geq 14$

Find triangulations for which these lower bounds are attained.

4. On the sphere it is possible to draw four countries each adjoining the other three; on the projective plane six countries can each adjoin the other five; and on the torus each of seven countries can adjoin the other six. The **problem of contiguous regions** on a given surface is the problem of finding the largest number of countries that can be drawn on that surface each adjoining all the others. The examples just cited show that the solution for the sphere, projective plane, and torus are at least 4, 6, and 7, respectively. The problem of contiguous regions bears on the coloring problem to the extent that the number of colors required on a surface is at least as large as

the largest number of mutually adjoining regions. However, maps can require *more* colors for their coloring than the number of mutually adjoining countries on the map. Find a map on the sphere requiring four colors in which no more than three countries are mutually adjoining. Thus the solution of the problem of contiguous regions does not solve the coloring problem.

5. Given n points on a surface, it may be possible to connect each pair of points by a curve on the surface so that no two curves intersect except at the given points. For example, in the plane or on a sphere this is possible with 2, 3, and 4 points but not with 5. The **thread problem** on a surface is the problem of determining the largest number of points on a surface that can be joined in this way. Show that the thread problem is equivalent to the problem of contiguous regions.

6. Suppose that five points on a sphere were each connected to the others by paths that intersect only at the given points. Explain that these paths divide the sphere into a complex for which $2E \geq 3F$, but that actually $V = 5$, $E = 10$, and $F = 7$. From this contradiction conclude that it is impossible that five points on the sphere be connected as described. Use this result to solve the thread problem on the sphere. Use the same technique to solve the thread problem on the projective plane and the torus.

7. Using the technique suggested by the preceeding exercise, show that N_χ is an upper bound for the solution of the thread problem on a surface of characteristic χ. How can this result be interpreted as evidence in favor of the Heawood conjecture?

REGULAR COMPLEXES

Given a surface $\mathscr{S}$, a regular complex on $\mathscr{S}$ is a complex $\mathscr{K}$ such that every face has the same number of vertexes and every vertex lies on the same number of faces. Regular complexes are a generalization of the regular polyhedra studied in §1. Examples include the torus in Figure 27.3 and the projective plane in Figure 27.2. To a certain extent they can be studied using the Euler characteristic just as regular polyhedra were studied earlier.

Exercises

8. Let $\mathscr{K}$ be a regular complex on a surface $\mathscr{S}$ of characteristic χ. Let a be the number of vertexes on each face, and let b be the number of faces at

each vertex. Show that

$$\frac{1}{a} + \frac{1}{b} = \frac{1}{2} + \frac{\chi}{2E} \qquad (2)$$

9. A regular complex is **degenerate** if either a or b is 2. Show that only the sphere and projective plane can have degenerate regular polyhedra.

10. Consider a nondegenerate regular complex on a projective plane. Show that $3 \leq a, b \leq 6$, and find all solutions to the Diophantine equation (2). For each solution determine whether or not there actually is a regular complex on the projective plane for those values of a and b.

11. Solve the equation (2) for the case of a regular complex on torus. Show that there are an *infinite* number of regular complexes on the torus for each solution of (2). Show that the Klein bottle also has an infinite number of regular complexes.

12. For surfaces with negative Euler characteristic, show that $3 \leq a$, $b < 6(1 - \chi)$ and if $F > N_\chi$, then $a, b < N_\chi$.

13. Use the results of Exercise 12 to investigate regular complexes on the surfaces of negative Euler characteristic.

---

§28 GRADIENT VECTOR FIELDS

Let $\mathscr{S}$ be a compact surface in space, for example the lumpy sphere in Figure 28.1. If we choose a reference plane, then every point P on $\mathscr{S}$ lies a certain distance $h(P)$ above this plane. Under suitable assumptions regarding

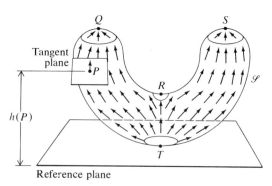

Figure 28.1   A lumpy sphere with vector field.

the surface $\mathscr{S}$ there exists a continuous vector field V on $\mathscr{S}$ such that at each point P the vector $V(P)$ lies in the tangent plane to $\mathscr{S}$ at P and points in the direction of greatest increase of h (i.e., along the path of steepest ascent on $\mathscr{S}$). V is called the **gradient vector field** of h. The precise nature of the assumptions required to guarantee the existence of the gradient vector field is not important here. They must ensure that $\mathscr{S}$ is smooth enough to have a tangent plane at every point that turns continuously as the point P moves across $\mathscr{S}$. They belong properly to the subject of differential geometry. We intend to study the topological aspects of V, assuming without proof the basic properties that flow from the smoothness of $\mathscr{S}$. Eventually we shall study tangent vector fields in general under the same circumstances (in §31). The study of gradient vector fields is a necessary preliminary. To the extent that our results depend on facts from differential geometry, our proofs are incomplete. However, new proofs of the main theorems from an entirely different point of view will appear in Chapter Six.

As usual with vector fields, interest centers on the **critical points**, points where $V(P)$ is zero (points Q, R, S, and T in Figure 28.1). For gradient vector fields these are the points where the tangent plane is parallel to the reference plane. We will assume that the critical points are *isolated*; that is, each has a neighborhood in which it is the only critical point. Then it follows from the compactness of $\mathscr{S}$ that there are only a finite number of critical points. As in our earlier investigation of gradient vector fields, these critical points represent only a few types of critical points. There are only *nodes* at the local maxima and minima on $\mathscr{S}$ (the points Q, S, and T in Figure 28.1) and *cross points* of various types (R in Figure 28.1).

To find the index of a critical point P, we choose a neighborhood of P in which P is the only critical point and project this neighborhood together with the vectors V onto the tangent plane at P. Figure 28.2 shows this process applied to the points Q, R, and T from Figure 28.1. The projected vectors form a continuous vector field in the tangent plane. We define the index of V to be the index of this plane vector field at P computed as in Chapter Two. The notion of index leads us to the following amazing theorem.

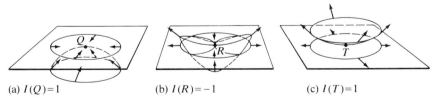

(a) $I(Q)=1$ (b) $I(R)=-1$ (c) $I(T)=1$

Figure 28.2   Finding the index. (a) $I(Q) = 1$. (b) $I(R) = -1$. (c) $I(T) = 1$.

Theorem

Let the points $P_1, P_2, \ldots, P_n$ *be the critical points of a gradient vector field on the compact surface* $\mathscr{S}$. *Then*

$$I(P_1) + I(P_2) + \cdots + I(P_n) = \chi(\mathscr{S})$$

Note that this conclusion is borne out by our example. There we have three nodes and one cross point, so the sum of the indexes of the critical points is 2, which is the Euler characteristic of the sphere. This theorem demonstrates the importance of the Euler characteristic by showing again how it is involved with concrete geometric properties of surfaces.

For the proof, which is based on some ideas of Banchoff [2], we will use the formula for the index

$$I(P) = 1 + \tfrac{1}{2}(e - h)$$

developed in Exercise 9 of §9 in which e and h are the numbers of elliptic and hyperbolic sectors at P. In the present context this becomes

$$I(P) = 1 - \frac{h}{2}$$

since there are no elliptic sectors. In order to find a simple method for counting the hyperbolic sectors, consider the example of Figure 28.2b. The point R is a saddle point, and the figure shows the four separatrixes that mark off the four hyperbolic sectors. Note that in each sector the surface crosses the tangent plane. It can be proven that, if $\mathscr{S}$ is smooth enough, then every sufficiently small circle drawn around a critical point P will intersect the tangent plane exactly h times, where h is the number of hyperbolic sectors at P. Note that the behavior of our example at the points Q and T bears out this assertion.

To connect these observations with homology, let $\mathscr{K}$ be a triangulation of $\mathscr{S}$. Among the vertexes of $\mathscr{K}$ we include all the critical points $P_1, P_2, \ldots, P_n$ of V and among the edges we include segments making up a small circle about each critical point. Figure 28.3 displays such a triangulation of the lumpy sphere.

Definition

A vertex P of a triangle is called **middle** *for that triangle when $h(P)$ is between the heights of the other two vertexes of the triangle.*

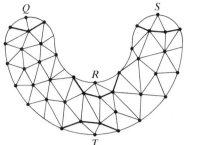

Figure 28.3

Consider the application of this definition to a critical point P. The point P lies in a number of triangles each of which includes one side from a small circle about P. If in such a triangle the side opposite P intersects the tangent plane at P, then P will be middle in that triangle. Thus the number of hyperbolic sectors at P equals the number of triangles in the triangulation for which P is middle. In other words,

$$I(P) = 1 - \tfrac{1}{2}(\text{number of triangles for which } P \text{ is middle}) \qquad (1)$$

Note that if this formula is used at an ordinary point, the result is zero, as it should be, since an ordinary point is like a point with two hyperbolic sectors.

To complete the proof, assume that no triangle contains two vertexes of the same height. This can always be arranged by a slight distortion of the surface if necessary. Then every triangle has a middle vertex. Now

$$I(P_1) + I(P_2) + \cdots + I(P_n) = \sum_{\substack{\text{critical} \\ \text{points} \\ P}} I(P)$$

$$= \sum_{\substack{\text{critical} \\ \text{points} \\ P}} [1 - \tfrac{1}{2}(\text{number of triangles for which } P \text{ is middle})]$$

$$= \sum_{\substack{\text{all} \\ \text{vertexes} \\ P}} [1 - \tfrac{1}{2}(\text{number of triangles for which } P \text{ is middle})]$$

$$= V - \tfrac{1}{2}F$$

Since for triangulations $3F = 2E$, $F = 2E - 2F$, it follows that

$$I(P_1) + I(P_2) + \cdots + I(P_n) = V - \tfrac{1}{2}(2E - 2F)$$
$$= V - E + F = \chi(\mathscr{S})$$

Exercises

1. Verify the conclusion of the theorem of this section for the following surfaces by labeling all critical points and finding the sum of their indexes.

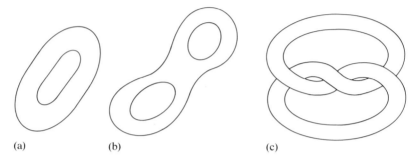

(a)   (b)   (c)

2. Draw examples of surfaces to fit the following specifications:

(a) a sphere with a monkey saddle (modified bowling ball)

(b) a torus with three cross points

(c) a surface with no cross points

3. Try to find out whether the theorem of this section applies to surfaces that *interpenetrate*, that is, intersect themselves, by studying some examples. Start with the Klein bottle in Figure 19.12d.

THE EULER CHARACTERISTIC
FOR THREE-DIMENSIONAL MANIFOLDS

In §23 we introduced three-dimensional complexes—topological spaces made by topological identification of polyhedral solids along faces—and three-dimensional manifolds—three-dimensional complexes in which each point has a neighborhood equivalent to a solid sphere. We intend to determine the Euler characteristic of these manifolds. Our discussion is based on Blackett [4].

In order to understand the problems involved in constructing manifolds, consider the following example. The complex consists of a single cube identified at opposite faces rather like the example in Figure 23.7, only this time each face is rotated 180° before identification. The result is shown in

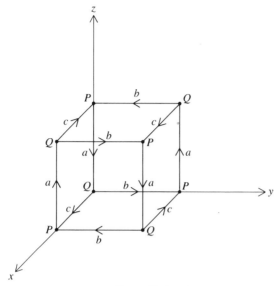

Figure 28.4

Figure 28.4. In order to be perfectly explicit about the rotation, we stipulate the following: the face of the cube where $x = 0$ is rotated about the z-axis $180°$ and then identified with the face at $x = 1$; the face at $y = 0$ is rotated about the x-axis before identification to the face at $y = 1$; and the face at $z = 0$ is rotated about the y-axis before being identified to the face at $z = 1$. This leads to the vertex and edge identifications shown in Figure 28.4. The resulting complex $\mathcal{K}$ might be regarded as a three-dimensional analogue of the Klein bottle. In any event, $\mathcal{K}$ is some sort of three-dimensional twisted torus.

It might be well to remark here that the entire theory of homology generalizes easily to three-dimensional complexes (indeed to n-dimensional complexes). The only difference is that there is one more of everything: one more chain group, one more boundary operator, one more group of cycles and group of boundaries, and of course, one more homology group and one more Betti number. The Euler characteristic of a complex is

$$\chi = V - E + F - S$$

where V, E, F, and S are the numbers of vertexes, edges, faces, and polyhedral solids in the complex. In our example $S = 1, F = 3, E = 3, V = 2$ (on account of the identifications); hence $\chi = 2 - 3 + 3 - 1 = 1$, suggesting that perhaps

$\mathscr{K}$ is really a sort of three-dimensional projective plane. As in two dimensions, an invariance theorem assures us that the Euler characteristic represents a topological property of the complex. However, we have not even verified that $\mathscr{K}$ is a manifold!

Let us examine $\mathscr{K}$ to see whether it is a manifold or not. There are four types of points to consider: points in the interior of the cube, points in the interior of a face, points in the interior of an edge, and vertexes. In each case the question is; Does the point have a neighborhood equivalent to a solid sphere? Figure 28.5a,b, and c shows neighborhoods of the first three types of points. All these neighborhoods are equivalent to solid spheres. In the case of the third type of point, the neighborhood consists of four quarter spheres. Turning to the last category of points (Figure 28.6), a neighborhood is a union of four *corner solids*. This union is *not* a solid sphere. We argue as

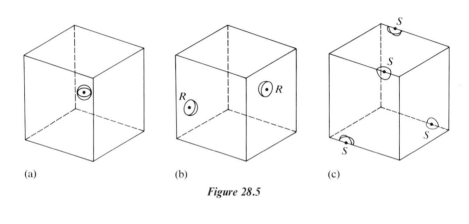

(a)   (b)   (c)

Figure 28.5

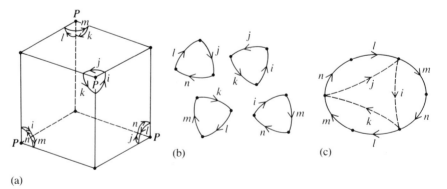

(a)

(b)   (c)

Figure 28.6

follows: if it were a solid sphere, then the boundary of that solid sphere would be equivalent to a spherical surface. This boundary is composed of the four triangles shown in Figure 28.6b, the bases of the corner solids at the vertex P. However, when these triangles are sewn together, the result is not a sphere but a projective plane (Figure 28.6c). Therefore the neighborhood at P is not a spherical solid, but some sort of solid bounded by a projective plane! It follows that the complex $\mathcal{K}$ is not a manifold at all, there being just two points P and Q that fail to have neighborhoods equivalent to a solid sphere.

This example demonstrates one difficulty arising in the study of three-manifolds: that of finding examples of three complexes that are three-manifolds. The following unfortunate theorem exposes another difficulty.

Theorem

If $\mathcal{K}$ is a three-dimensional manifold, then the Euler characteristic of $\mathcal{K}$ is zero.

It follows from this theorem that the Euler characteristic can tell us *nothing* about differences among three-dimensional manifolds. What a change from the situation with regard to two-dimensional manifolds (surfaces). There the Euler characteristic can assume any integer value less than or equal to two, and all of our applications have exploited the value of the characteristic! In contrast, according to this theorem, the characteristic carries no information about three-dimensional manifolds. (This vanishing of the Euler characteristic actually occurs for all odd-dimensional manifolds.)

The proof of the theorem will be given in a series of exercises. Let $\mathcal{K}$ be a three-dimensional manifold. Let V be the number of vertexes, E the number of edges, F the number of faces, and S the number of polyhedral solids of $\mathcal{K}$, where identifications are taken into account. Let V', E', and S' be the corresponding numbers *if identifications are ignored*. (For the example of Figure 28.4, $V' = 8$, $E' = 12$, $F' = 6$, and $S' = 1$.)

Let $P_1, P_2, \ldots, P_n$ be the vertexes of $\mathcal{K}$ ($n = V$). At each vertex P_i choose a neighborhood equivalent to a solid sphere. This neighborhood, like the neighborhood of the vertex P shown in Figure 28.6a, will be composed of a certain number of corner solids. The boundary of the neighborhood will be a two-dimensional complex equivalent to a sphere. Let V_i, E_i, and F_i be the number of vertexes, edges, and faces in this boundary complex taking identifications into account. (For the vertex P in Figure 28.6, $V = 6$, $E = 9$, and $F = 4$, and $V - E + F = 1$, as it should be for a projective plane.)

Exercises

4. Explain why

$$V_1 - E_1 + F_1 = 2$$
$$V_2 - E_2 + F_2 = 2$$
$$\vdots$$
$$V_n - E_n + F_n = 2$$

5. Explain why

$$2E = V_1 + V_2 + \cdots + V_n$$

and

$$V' = F_1 + F_2 + \cdots + F_n$$

6. At each vertex P_i let V_i' and E_i' be the number of vertexes and edges in the spherical boundary complex, *ignoring identifications.* Explain why $V_i' = E_i' = 2E_i$, and

$$2E' = V_1' + V_2' + \cdots + V_n'$$

in order to deduce

$$E' = E_1 + E_2 + \cdots + E_n$$

7. From the preceding exercises deduce

$$2E - E' + V' = 2V \qquad (2)$$

Explain why $S' = S$, $F' = 2F$, and

$$V' - E' + F' = 2S'$$

in order to deduce

$$2F - E' + V' = 2S \qquad (3)$$

From (2) and (3) it follows immediately that

$$V - E + F - S = 0$$

8. Compute the Euler characteristic of the topological space of Figure 28.4, Figure 23.7, and the two spaces in Exercise 12 of §23.

9. Find a scheme of twisted identifications for the faces of a cube that does lead to a manifold. Use the Euler characteristic to help reject nonmanifolds.

§29 INTEGRAL HOMOLOGY

The homology developed so far, called **homology (mod 2)**, is unsatisfactory in some ways. For example, homology (mod 2) is unable to distinguish the torus from the Klein bottle; these two surfaces have the same homology (mod 2). This section introduces a more powerful type of homology, **integral homology**, which will carry more information about the topological space to which it is applied. Among other things, it will distinguish the torus from the Klein bottle.

Integral homology begins, like homology (mod 2), with a topological space divided into simplexes so as to form a complex $\mathcal{K}$. The extra information to be incorporated into integral homology comes from directing the complex.

Definition

*The complex $\mathcal{K}$ is **directed** when each edge of $\mathcal{K}$ is given a definite direction from one endpoint to the other, and each polygon is given a definite direction around the polygon.*

Figure 29.1 gives two examples of directed complexes. The first is a torus divided into four triangles; the second is a plane model of the Klein bottle. Directions are indicated by arrows along each edge and inside each polygon.

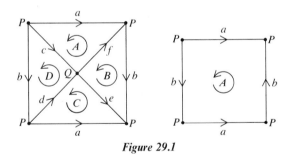

Figure 29.1

For simplicity we have directed all polygons clockwise, but the choice is arbitrary. Any direction is permitted for any face, the important thing is to choose *some* direction for each edge and each face.

Information concerning the direction of the complex is incorporated into the incidence coefficients. The best way to explain this is by example. Here are the incidence coefficients for the two examples of Figure 29.1. For the torus:

	a	b	c	d	e	f
P	0	0	-1	1	1	-1
Q	0	0	1	-1	-1	1

	A	B	C	D
a	1	0	-1	0
b	0	1	0	-1
c	-1	0	0	1
d	-1	1	0	0
e	0	-1	1	0
f	0	0	1	-1

and for the Klein bottle:

	a	b
P	0	0

	A
a	0
b	-2

You will immediately notice the *negative* incidence coefficients. Negative coefficients occur (1) between a point and an edge when the point is the *initial* point of the edge and (2) between an edge and a face when the direction of the edge *opposes* the direction of the face. For example, on the torus the incidence coefficient of P on the edge c is -1 while the incidence coefficient of Q on c is $+1$. The incidence coefficient of P on a is 0 because P is incident twice with the edge a, once as initial vertex and once as final vertex for a total incidence of $-1 + 1 = 0$. Searching for examples of (2), we find on the torus that the incidence coefficient of the edge c with the face A is -1, while the incidence coefficient of c with the face D is $+1$. Turning to the Klein bottle, notice another new feature: while the edges a and b both appear twice on the boundary of the face A, the incidence coefficient of a on A is 0 since a appears once in agreement with and once opposed to the direction of A, but the incidence coefficient of b on A is -2 since b is both times opposed to the direction of A. Thus these new incidence coefficients distinguish toroidal from twisted pairs of edges. With these examples in mind, you should have no trouble justifying the other incidence coefficients in Figure 29.1.

Exercises

1. Assign directions to the simplexes of the two complexes $\mathscr{K}$ and $\mathscr{L}$ of §23. Compute the incidence coefficients of the resulting directed complexes, and compare them with the incidence coefficients given in §23. In the case of the torus, compare the incidence coefficients with those of the Klein bottle given above.

2. Describe incidence coefficients for the normal forms of other compact, connected surfaces.

3. Give a formal definition for incidence coefficients on a directed complex.

The chains for integral homology are defined as abstract sums of simplexes with integer coefficients. As examples from the torus of Figure 29.1 we have

$$2A - B \qquad \text{(a 2-chain)}$$
$$2a + b - 2c \qquad \text{(a 1-chain)}$$
$$3Q - 3P \qquad \text{(a 0-chain)}$$

The plus and minus signs in these chains have only a formal significance. In other words, they serve to connect the individual terms together but do not represent usual arithmetic operations with numbers, or even the union of sets of simplexes as with homology (mod 2). These chains may be regarded as simply a list of simplexes in which each simplex is assigned an integer as coefficient. The addition of such chains, however, does follow the usual rules for the addition of integers. Thus if $a + c - 17e$ is another 1-chain, then $(2a + b - 2c) + (a + c - 17e) = 3a + b - c - 17e$. The algebra of integral chains is easier in some ways than the algebra of chains (mod 2) since it is closer to the familiar algebra of the integers. The price of this simplification is the loss of intuitive geometric meaning. Here is the formal definition of the integral chains.

Definition

Let $\mathscr{K}$ be a directed complex, and let $S_1, S_2, \ldots, S_n$ be the k-simplexes of $\mathscr{K}$ ($k = 0, 1, 2$). An **integral k-chain** *of $\mathscr{K}$ is a sum*

$$C = a_1 S_1 + a_2 S_2 + \cdots + a_n S_n$$

where the coefficients $a_1, a_2, \ldots, a_n$ can be any integers positive negative or zero. If

$$C' = b_1 S_1 + b_2 S_2 + \cdots + b_n S_n$$

is a second k-chain, then the sum $C + C'$ is defined by

$$C + C' = (a_1 + b_1)S_1 + (a_2 + b_2)S_2 + \cdots + (a_n + b_n)S_n$$

With this operation the set $C_k(\mathcal{K})$ of integral k-chains is a group whose zero is the chain for which $a_1 = a_2 = \cdots = a_n = 0$. The inverse of the k-chain C is given by

$$-C = -a_1 S_1 - a_2 S_2 - \cdots - a_n S_n$$

The chain group $C_k(\mathcal{K})$ is infinite.

What is one to think of these chains? The chains (mod 2) can also be written as sums $a_1 S_1 + a_2 S_2 + \cdots + a_n S_n$, but the coefficients here are either zero or one, signifying whether or not the simplex is in the chain. Therefore the chains (mod 2) have a simple geometric interpretation as sets of simplexes and can be easily visualized. Unfortunately integral chains have no similar intuitive geometric interpretation (except the chain $-S$, where S is a simplex, which can be regarded as the simplex S with the reverse direction). What intuitive meaning can be ascribed to the integral chain groups is described as follows: just as homology (mod 2) provides an algebra for handling certain simple counting operations (where only the evenness or the oddness of the result is important), so integral homology provides an algebra for handling simple counting operations without the peculiar restriction to evenness and oddness. The 1-chain $2a + b - 2c - 3d + e$ simply counts the edge a twice, the edge b once, the edge c minus two times, and so forth. Despite the lack of a simple geometric interpretation, the integral chains are a more powerful geometric tool than chains (mod 2). Although chains (mod 2) played a dominant role in the early history of combinatorial topology, they are now completely eclipsed by integral chains (and still more arcane types). In any event, no matter how it is justified, integral homology represents a step away from geometry toward abstract algebra. Much as we may regret it, this is a step well worth taking.

Definition

*The **boundary** of a k-simplex S ($k = 1, 2$) is the $(k-1)$-chain whose co-efficients are the incidence coefficients of S with each $(k-1)$-simplex. Then* *the boundary of the k-chain $a_1S_1 + a_2S_2 + \cdots + a_nS_n$ is defined by additivity to be*

$$\partial(a_1S_1 + a_2S_2 + \cdots + a_nS_n) = a_1\partial(S_1) + a_2\partial(S_2) + \cdots + a_n\partial(S_n)$$

As examples, on the torus of Figure 29.1 we have $\partial(A) = a - c - d$, and $\partial(c) = Q - P$. You should also verify that $\partial(2A - B) = 2a + b - 2c - 3d + e$, $\partial(2a + b - 2c - 3d + e) = 0$ and $\partial(c - e + f) = 3Q - 3P$.

Definition

*A k-chain C is a **k-cycle** if $\partial(C) = 0$ $(k = 1, 2)$. A k-chain C is a **k-boundary** if there exists a $(k + 1)$-chain C' such that $C = \partial(C')$ $(k = 0, 1)$. By convention all 0-chains are considered 0-cycles, while the only 2-boundary is the zero 2-chain. The sets $Z_k(\mathcal{K})$ and $B_k(\mathcal{K})$ of k-cycles and k-boundaries are subgroups of $C_k(\mathcal{K})$ satisfying the inclusion relation $B_k \subseteq Z_k \subseteq C_k$; in other words, every boundary is a cycle.*

This definition is just as for homology (mod 2). It is already possible through these notions to distinguish the homology of the torus from that of the Klein bottle. The torus has a nonzero 2-cycle, $\partial(A + B + C + D) = 0$; however, the Klein bottle has no nonzero 2-cycle, since $\partial(A) = -2b$.

Definition

*Two k-chains C_1 and C_2 are **homologous**, written $C_1 \sim C_2$, if they differ by a boundary, that is, if $C_1 - C_2$ is a k-boundary or if C_1 equals C_2 plus a k-boundary. The relation of homology enjoys all the usual algebraic properties of equality.*

The use of subtraction in the definition of homology is a slight departure from the tradition established in previous sections, but it is appropriate to integral homology. The plus sign was possible formerly only because the groups involved were idemgroups.

Definition

*The group Z_k of k-cycles remains a group when homology replaces equality. The resulting group is called the kth **homology group** of $\mathcal{K}$ and is notated $H_k = Z_k/B_k$.*

Exercises

4. Verify that C_k satisfies all the group axioms (§23).

5. Verify that the boundary operator ∂ is additive.

6. Prove that $B_k \subseteq C_k$ ($k = 0, 1, 2$), and verify the algebraic properties of homology.

7. If homology were defined with a plus sign instead of a minus sign, show that it would have only one of the basic algebraic properties of equality.

Are the integral homology groups *invariant*? You will recognize immediately the importance of this question. The construction of the integral homology groups of a topological space depends on choosing a complex $\mathcal{K}$, equivalent to the space, directing the complex, setting up incidence coefficients, etc. Unless the homology groups are independent of the choices involved in this construction, they will not reflect topological properties of the original topological space. Luckily this independence or invariance can be established by arguments similar to those used in §25. The new feature is the choice of a direction for the simplexes of the complex $\mathcal{K}$. It is easy to show, however, that this choice has no effect on the chain groups, and hence no effect on the homology groups. The reason is this: for each simplex S, the chain groups contain both S and $-S$; that is, they contain an element for each possible direction of S. A change in direction simply replaces S by $-S$, an operation that is clearly a group isomorphism. Thus the chain groups corresponding to different directions of $\mathcal{K}$ are equivalent. With this preliminary step disposed of, one can easily complete the proof of the following theorem.

Invariance Theorem

The integral homology groups of a compact surface are independent of the complex used for their construction provided the complex can be reduced to the plane model of the surface using cutting and pasting operations. More generally (using the Hauptvermutung), the integral homology groups of any triangulable space of dimension two are independent of the choice of triangulation.

EXAMPLES OF INTEGRAL HOMOLOGY GROUPS

One of the consequences of the invariance theorem is that the homology groups of surfaces can be computed from the plane models. For the torus (Figure 23.4), no matter how it is directed, all incidence coefficients are zero. Therefore there are no boundaries other than $\varnothing$, and every chain is a cycle. Thus homology is the same as equality, and $H_k = C_k$. The zeroth homology group consists of the chains nP, where P is the lone vertex of the torus and n is any integer. Clearly the mapping $nP \to n$ is an isomorphism of H_0 onto the group of integers $\mathscr{Z}$. On similar grounds, H_2, which consists of all the chains nA, where A is the one polygon of the torus, is also isomorphic to $\mathscr{Z}$. The first homology group consists of all chains $na + mb$, where both n and m are integers. This group is isomorphic to the two-dimensional group of integers $\mathscr{Z}^2$. Thus we have $H_0 \cong \mathscr{Z}$, $H_1 \cong \mathscr{Z}^2$, and $H_2 \cong \mathscr{Z}$.

As a second example, consider the Klein bottle (Figure 29.1). We have already noted that the Klein bottle has no 2-cycle. Instead there is a 1-boundary, since $\partial(A) = -2b$. This is the only difference between the homology of the Klein bottle and the homology of the torus. Zero-dimensional homology is not affected, so $H_0 = \mathscr{Z}$, as with the torus. Since $Z_2 \cong \mathscr{C}_1$, $H_2 \cong \mathscr{C}_1$. This disposes of the second homology group. Turning to the first homology group, note that every 1-chain is a 1-cycle, so that like the torus, the group of 1-cycles consists of all sums $na + mb$, where n and m are integers. The group B_1 contains, in addition to $-2b$, all integer multiples of $-2b$. Thus B_1 contains all *even* multiples of b. Therefore in the homology group H_1, $mb \sim \varnothing$ for every even integer m. At the same time, if m is odd, $m = 2p + 1$, then $mb = 2pb + b \sim b$. In other words, the subgroup of H_1 that arises from the cycles mb consists in reality of just two cycles, $\varnothing$ and b, to which all the others are homologous. This subgroup is isomorphic to $\mathscr{C}_2$. H_1 also contains the elements na, none of which are homologous to each other, so that the subgroup consisting of these elements is isomorphic to $\mathscr{Z}$. The situation encountered here is described in the following definition.

Definition

Let the group G contain the subgroups G_1 and G_2. Supposing that the only element common to G_1 and G_2 is the zero, and every element of G is the sum of an element from G_1 plus an element from G_2, then G is called the **direct sum** *of these subgroups, written $G = G_1 \oplus G_2$.*

Returning to the Klein bottle, let A and B stand for the subgroups of H_1 consisting of the chains of the forms na and mb, respectively. Then obviously

$H_1 = A \oplus B$, while $A \cong \mathscr{Z}$ and $B \cong \mathscr{C}_2$. Thus for the Klein bottle we have found that $H_0 \cong \mathscr{Z}$, $H_1 \cong \mathscr{Z} \oplus \mathscr{C}_2$, and $H_2 \cong \mathscr{C}_1$.

Exercises

8. Prove an invariance lemma (as in §25) for integral homology, and use it to prove the invariance theorem.

9. Show that $\mathscr{D}_2 \cong \mathscr{C}_2 \oplus \mathscr{C}_2$ and $\mathscr{Z}^2 \cong \mathscr{Z} \oplus \mathscr{Z}$.

10. Compute the integral homology groups for the sphere, projective plane, double torus, and connected sum of three projective planes.

§30 TORSION AND ORIENTABILITY

In this section we investigate the algebraic structure of the integral homology groups. As with homology (mod 2), this is a necessary preliminary in order to discover the topological significance of these groups. This time we use a notion of dependence appropriate to infinite groups.

Definition

*Let G be a group and A be a subset of G. An element x is **dependent on** A if there exist integers $a, a_1, a_2, \ldots, a_n$ and elements $x_1, x_2, \ldots, x_n$ in A such that*

$$ax = a_1 x_1 + a_2 x_2 + \cdots + a_n x_n$$

*where $a \neq 0$. Otherwise x is called **independent of** A. The subset A **spans** G if every element of G is dependent on A. The subset A is **independent** if no element of A depends on the other elements of A. An independent spanning set is called a **basis** for G.*

There is no possibility of confusing this notion of dependence with that defined in §26, since the latter will not be used again. The present notion has all the properties of the former, the proofs being analogous to those of §26.

Fundamental Properties of Dependence

(a) Every element of A depends on the set A.

(b) *If every element of A depends on a second set B, then every element depending on A also depends on B.*

(c) *If x depends on the set* $\{y_1, y_2, \ldots, y_n\}$ *but does not depend on the set* $\{y_1, y_2, \ldots, y_{n-1}\}$, *then* y_n *depends on the set* $\{y_1, y_2, \ldots, y_{n-1}, x\}$.

One new feature of this theory of dependence is that the groups being studied are infinite and hence need not be spanned by a finite set. The group G is called **finitely generated** if there exists a finite subset $\{x_1, x_2, \ldots, x_n\}$ of G such that every element of G is equal to a linear combination $a_1 x_1 + \cdots + a_n x_n$. A finitely generated group clearly has a finite spanning set. For these groups the theorems of §26 are easily generalized.

Theorem 1

Let G be finitely generated and let A be a finite set spanning G. Then A contains a finite subset that is a basis for G.

Theorem 2

Let G be finitely generated. Then any two finite bases for G contain the same number of elements. This number is called the **rank** *of G.*

Theorem 3

Let G and H be finitely generated groups, and let f be a homomorphism from G to H. The rank of the kernel $K(f)$ is called the **nullity** *of f, while the rank of the range $R(f)$ is called the* **rank** *of f. The rank of f plus the nullity of f equals the rank of G.*

Theorem 4

Let H be a finitely generated subgroup of the finitely generated group G. Then the quotient group G/H is also finitely generated, and the rank of G/H equals the rank of G minus the rank of H.

Assuming that the groups involved are finitely generated, given a complex $\mathcal{K}$, the ranks h_0, h_1, and h_2 of the integral homology groups are called the **integral Betti numbers** of $\mathcal{K}$. The integral Betti numbers need *not* be the same as the Betti numbers defined in §26. For example, for the Klein bottle $h_0 = r(\mathcal{Z}) = 1$, $h_1 = r(\mathcal{Z} \oplus \mathcal{C}_2) = 1$ and $h_0 = r(\mathcal{C}_1) = 0$, as compared with the values $h_0 = 1$, $h_1 = 2$, $h_0 = 1$ from §26.

Exercises

1. Supply proofs of the fundamental properties of independence and Theorems 1–4 by imitating the proofs of the corresponding theorems from §26.

2. Prove that the chain groups C_0, C_1, and C_2 of a complex $\mathscr{K}$ are finitely generated if and only if $\mathscr{K}$ is compact. Show that in case $\mathscr{K}$ is compact, the ranks c_0, c_1, and c_2 of these groups are the numbers of simplexes in $\mathscr{K}$.

3. Let G be the direct sum of finitely generated groups H_1 and H_2. Prove that G is finitely generated and that the rank of G is the sum of the rank of H_1 and the rank of H_2.

4. Show that $r(\mathscr{C}_1) = r(\mathscr{C}_2) = \cdots = r(\mathscr{C}_n) = 0$.

5. Find the integral Betti numbers for the sphere, torus, projective plane, Klein bottle, double torus, and the connected sum of three projective planes.

6. Prove the Euler-Poincaré formula for the integral Betti numbers, $\chi = c_0 - c_1 + c_2 = h_0 - h_1 + h_2$. Check that the formula holds for the surfaces mentioned in the previous exercise.

TORSION

The rank of a finitely generated group, although important in describing the group structure, by no means tells the whole story.

Definition

*Let G be a group. The **torsion subgroup** of G is the subset T of elements x in G for which there is an integer $n \neq 0$ such that $nx = 0$. If $T = G$, then G is called a **torsion group**.*

The elements x of the torsion subgroup all depend on zero, and the torsion subgroup itself has rank zero. All finite groups are torsion groups. The idemgroups in particular are torsion groups; for them n may always be taken as two (in the above definition). The torsion group contributes nothing toward the rank of G. More precisely, we have the following theorem.

Direct Sum Theorem

Every finitely generated group G is the direct sum of its torsion group T and a group $\mathscr{Z}^r$, where r is the rank of G.

The proof of this theorem depends on two lemmas.

Lemma 1

Let f be a homomorphism from a group G onto a group $\mathscr{Z}^r$. Let T be the kernel of f. Then $G = T \oplus \mathscr{Z}^r$.

For the proof let $e_1, e_2, \ldots, e_r$ be the standard basis for $\mathscr{Z}^r$; that is, $e_1 = (1, 0, 0, \ldots), e_2 = (0, 1, 0, \ldots), \ldots,$ and $e_r = (0, 0, \ldots, 0, 1)$. Now choose elements $\{x_1, x_2, \ldots, x_r\}$ in G such that $f(x_1) = e_1, f(x_2) = e_2$, and so forth. Let B be the subgroup of G consisting of linear combinations $a_1 x_1 + a_2 x_2 + \cdots + a_r x_r$. The homomorphism f restricted to B is one to one. To prove this, let $s = a_1 x_1 + \cdots + a_r x_r$ be an element of B. If $f(x) = 0$, then $0 = f(a_1 x_1 + \cdots + a_r x_r) = a_1 e_1 + \cdots + a_r e_r = (a_1, a_2, \ldots, a_r)$. It follows that $a_1 = a_2 = \cdots = a_r = 0$; hence $x = 0$. This proves that B is isomorphic to $\mathscr{Z}^r$. The proof of the lemma is concluded by proving that $G = T \oplus B$. (1) Let x be in $T \cap B$. Then $x = a_1 x_1 + \cdots + a_r x_r$ for some integers $a_1, \ldots, a_r$ and $f(x) = 0$. By the argument just given, it follows that $x = 0$. (2) Let x be any element in G. Then $f(x) = a_1 e_1 + \cdots + a_r e_r$ for some integers $a_1, \ldots, a_r$. Let $b = a_1 x_1 + \cdots + a_r x_r$, and let $t = x - b$. Then $b \in B, t \in T$, and $x = t + b$. This completes the proof.

Lemma 2

Every subgroup of $\mathscr{Z}^n$ is of the form $\mathscr{Z}^m$, where $m \leq n$.

The proof is by induction on n. Consider first a subgroup B of $\mathscr{Z}^1 = \mathscr{Z}$. If B contains any element of $\mathscr{Z}$ but zero, then B contains some positive elements. Let b be the smallest positive element in B. We shall prove that B consists entirely of the elements kb (where $k \in \mathscr{Z}$) and so B is isomorphic either to $\mathscr{Z}^0 = \{0\}$ or to $\mathscr{Z}$. Let a be any other element of B. By dividing a by b, we find a quotient k and remainder r, both integers, such that $a = kb + r$, where $0 \leq r < b$. From $r = a - kb$ it follows that r is in B. Since b is the smallest positive element of B, $r = 0$, whence $a = kb$.

Now assume the lemma is true for groups of the form $\mathscr{Z}^{n-1}$, and let B be a subgroup of $\mathscr{Z}^n$. Let $e_1, e_2, \ldots, e_n$ be the standard basis for $\mathscr{Z}^n$, and let f be the homomorphism from $\mathscr{Z}^n$ to $\mathscr{Z}$ defined by $f(a_1 e_1 + \cdots + a_n e_n) = a_1$. Consider f restricted to B, and let T be the kernel of this restricted homomorphism. There are two cases. If $T = B$, then B is entirely contained in the subgroup $\mathscr{Z}^{n-1}$ of $\mathscr{Z}^n$ spanned by $e_2, e_3, \ldots, e_n$. In this case we may apply the induction hypothesis directly to B since B is a subgroup of a group of the form $\mathscr{Z}^{n-1}$. Otherwise $T \neq B$ and so f maps B onto some range subgroup of

$\mathscr{Z}$ that by the previous paragraph is itself isomorphic to $\mathscr{Z}$. Then by lemma 1, $B = T \oplus \mathscr{Z}$, where T is a subgroup of $\mathscr{Z}^{n-1}$. The induction hypothesis can now be applied to T to complete the proof of the lemma.

Turning to the proof of the direct sum theorem, let $\{x_1, x_2, \ldots, x_r\}$ be a basis for G, and let B be the subgroup of G consisting of linear combinations $a_1 x_1 + \cdots + a_r x_r$. Just as in the proof of Lemma 1, B is isomorphic to $\mathscr{Z}^r$. Let $\{y_1, y_2, \ldots, y_n\}$ be a set of generators for G, that is, a set such that every element of G is equal to a linear combination $a_1 y_1 + \cdots + a_n y_n$. Since $\{x_1, x_2, \ldots, x_r\}$ is a basis for G, for each y there is an integer c_i such that

$$c_i y_i = a_1 x_1 + \cdots + a_r x_r$$

for some integers $a_1, \ldots, a_r$. In other words, $c_i y_i \in B$. Let e be the product of all the integers $c_1, c_2, \ldots, c_n$. Let f be the homomorphism $f(x) = ex$. By construction, the range of f is a subgroup of B. At the same time, the kernel of f is T (Exercise 9). By Lemma 2, the range of f is of the form $\mathscr{Z}^m$ for some $m \leq r$. By Lemma 1, $G = T \oplus \mathscr{Z}^m$. Since G contains a group of the form $\mathscr{Z}^r$, it follows that $m = r$. This concludes the proof.

The torsion subgroups of the integral homology groups are one means by which this homology distinguishes between the twisted and toroidal surfaces. The homology groups of the toroidal surfaces are **torsion free**; that is, they have no torsion elements, while the group H_1 for a twisted surface always contains some torsion.

The theorems proved in this section are part of the general theory of groups. It is clear that these algebraic results were inspired by their application to topology. The term "torsion group," for example, has a geometric meaning but no intrinsic algebraic significance. The development of group theory provides an example of the influence of one mathematical field on the content of another. Although subject to other influences as well, and useful in other fields of application, group theory and abstract algebra in general have grown hand in hand with topology in the twentieth century.

Exercises

7. Show that the torsion subgroup of a group *is* a subgroup!

8. Let f be a group homomorphism from G to H. Prove that f maps torsion elements from G to torsion elements in H.

9. Verify the statements about the homomorphism f made in the last step of the proof of the direct sum theorem.

10. Let $\mathscr{K}$ be a compact complex. Show that the groups C_k, Z_k, B_k are all of the form $\mathscr{Z}^m$ for some integer m. Conclude that H_k is finitely generated, and hence that $H_k = T_k \oplus \mathscr{Z}^p$, where p is the kth Betti number of $\mathscr{K}$.

11. Find the torsion groups and Betti numbers of all the compact surfaces, with and without boundary.

12. Find a complex $\mathscr{K}$ whose first homology group has torsion group $\mathscr{C}_3$. Do the same for the groups $\mathscr{C}_4$, $\mathscr{C}_5$, and $\mathscr{D}_2$.

ORIENTABILITY

This is a topological property, related to torsion, that provides another means for distinguishing the twisted from the toroidal surfaces homologically. Perhaps the best way to present it is through a fable.

FABLE

Once upon a time there was a topologist living on a cylinder (Figure 30.1), who was a specialist in winding numbers. As everyone knows, the most important thing in the study of winding numbers is to choose a direction of rotation as the direction of positive rotation. Thus the topologist always kept on hand a supply of little hoops, each with an arrow indicating the direction of positive orientation. During an investigation these would be handed out to friends assisting in the study. Whenever the topologist met a friend, they would check hoops. Always they were delighted to find that their hoops agreed on the direction of positive orientation.

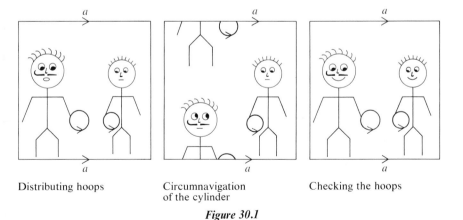

Distributing hoops   Circumnavigation   Checking the hoops
                              of the cylinder

Figure 30.1

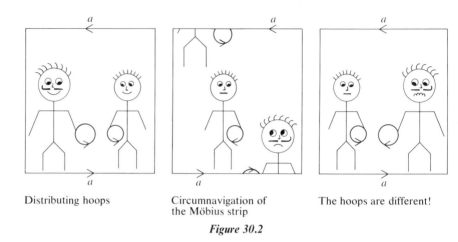

Distributing hoops   Circumnavigation of   The hoops are different!
                            the Möbius strip

Figure 30.2

Eventually the topologist grew successful enough to be invited to carry out a study of winding numbers on a nearby Möbius strip (Figure 30.2). Distributing hoops on the Möbius strip was an upsetting experience. The hoops agreed with each other only half the time. Becoming disoriented, the topologist, who despite repeated warnings insisted on drinking the local beer brewed from decayed hoops, soon returned home, abandoned topology, and lived happily ever after.

Moral. The Möbius strip is not orientable. This leads to the following definition.

Definition

Let $\mathcal{K}$ be a directed complex. The directions of two adjacent polygons A and B of $\mathcal{K}$ are said to **agree** *if the common edge e between the two polygons is oriented one way in the boundary of A and the other way in the boundary of B. In other words, e does not appear in $\partial(A + B)$ (see Figure 30.3). Otherwise the directions of A and B are said to* **disagree**. *The complex $\mathcal{K}$ is* **oriented** *if $\mathcal{K}$ is directed in such a way that directions of adjoining polygons always agree. A surface $\mathcal{S}$ is* **orientable** *if every complex equivalent to $\mathcal{S}$ can be oriented.*

Theorem

Let $\mathcal{S}$ be a compact, connected surface. If $\mathcal{S}$ is orientable, then $H_2(\mathcal{S}) \cong \mathcal{Z}$. If $\mathcal{S}$ is not orientable, then $H_2(\mathcal{S}) \cong \mathcal{C}_1$.

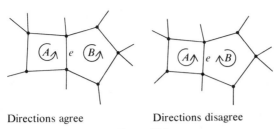

Directions agree   Directions disagree

Figure 30.3

For the proof, suppose first that $\mathscr{S}$ is orientable, and let $\mathscr{K}$ be a complex equivalent to $\mathscr{S}$. Let $C_1, C_2, \ldots, C_n$ be the polygons of $\mathscr{K}$ directed so that directions of adjacent polygons agree. Then $C_1 + C_2 + \cdots + C_n$ is a 2-cycle that is obviously not a 2-boundary, so $H_2(\mathscr{S}) \neq \mathscr{C}_1$. To complete the proof, we drop the assumption that $\mathscr{S}$ is orientable and assume instead that $H_2(\mathscr{S}) \neq \mathscr{C}_1$. We will now prove that $\mathscr{S}$ is orientable and that $H_2(\mathscr{S}) \cong \mathscr{L}$. Let $C = a_1C_1 + a_2C_2 + \cdots + a_nC_n$ be any 2-cycle not a 2-boundary. Thus at least one coefficient is not zero, say $a_1 = k \neq 0$. It follows that the co-efficients a_i of those polygons C_i adjacent to C_1 must all be $\pm k$ in order that C be a cycle. In turn, the coefficients of the polygons adjacent to these poly-gons must have coefficient $\pm k$. Since $\mathscr{S}$ is connected, we conclude, by con-tinuing this argument, that $C = \pm kC_1 \pm kC_2 \pm \cdots \pm kC_n$. Therefore $C' = C_1 \pm C_2 \pm \cdots \pm C_n$ (same signs as before) is also a 2-cycle. It is clear that $\mathscr{K}$ is orientable: an orientation is provided by redirecting the polygons preceeded by a minus sign in C', so that $C_1 + C_2 + \cdots + C_n$ becomes a 2-cycle instead of $C_1 \pm C_2 \pm \cdots \pm C_n$. We have also shown that every 2-cycle is a multiple of C'; thus $H_2(\mathscr{S}) \cong \mathscr{L}$.

Let $q = q(\mathscr{T})$ be some number defined for all topological spaces of some type $\mathscr{T}$. The number q is called a **topological invariant** if it is the same for topologically equivalent spaces; that is, if $\mathscr{T}$ and $\mathscr{T}'$ are topologically equivalent, then $q(\mathscr{T}) = (\mathscr{T}')$. In this book we have encountered several invariants for surfaces including the Euler characteristic χ and the Betti numbers h_0, h_1, h_2. A topological invariant always carries information concerning one or more topological properties. For example the Euler characteristic of a surface determines the sum of the indexes of the critical points on gradient vector fields and the map coloring number of the surface. For the Betti numbers we know that h_0 reflects the connectivity of the surface, and h_1 is related to the existence of closed curves on the surface that do not divide the surface. And now, according to the theory of this section, we see that h_2 reflects the orientability of the surface.

Exercises

13. Show directly from the definition that the cylinder is orientable while the Möbius strip is not.

14. Using the preceeding theorem, show that toroidal surfaces are orientable while twisted surfaces are not.

15. Among a class of topological spaces, a **complete set of invariants** is a list of invariants such that topologically distinct spaces in the class have distinct invariants. For the class of compact, connected surfaces, consider the following sets of invariants. In each case decide whether the set is complete or not. If the set is not complete, find two distinct surfaces with the same values for these invariants.

(a) $\{\chi, h_0\}$ 　　　　　　　　(b) $\{\chi, h_1\}$

(c) $\{\chi, h_2\}$ 　　　　　　　　(d) $\{h_0, h_1\}$

(e) $\{h_0, h_2\}$ 　　　　　　　　(f) $\{h_1, h_2\}$

(g) $\{h_0, h_1, h_2\}$

---

§31 THE POINCARÉ INDEX THEOREM AGAIN

This section continues the study of tangent vector fields and phase portraits begun in Chapter Two. Let $\mathscr{S}$ be a compact, connected surface imbedded smoothly enough in three-dimensional space that every point P on $\mathscr{S}$ has a tangent plane. A **tangent vector field** V is a function assigning to each point P of $\mathscr{S}$ a vector $V(P)$ tangent to $\mathscr{S}$ at P. Figure 31.1 gives an example on a sphere. We will assume that V is continuous and has only isolated critical points. The vector field in Figure 31.1, for example, has exactly six critical points. One of these, a saddle point P_5, is hidden in the back hemisphere. Associated with V, just as in the case of vector fields in the plane, is a family of paths on the surface $\mathscr{S}$, one path passing through each ordinary point of V, each path tangent to the vectors of V. These paths form the **phase portrait** of V. Some of these paths are shown in Figure 31.1a, and the entire phase portrait is presented on the normal form of $\mathscr{S}$ in Figure 31.1b. As in earlier discussions of vector fields, interest centers on the critical points where $V(P) = 0$. You may wish to review here the theory of critical points presented in Chapter Two, especially the computation of winding numbers and indexes of critical points. In the example in Figure 31.1 there are six critical points: 2 stable nodes, 2 unstable nodes, and 2 saddle points.

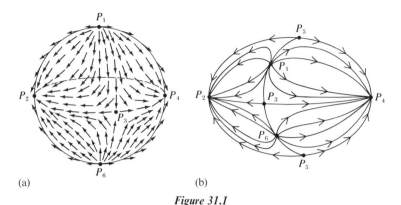

(a)   (b)

Figure 31.1

In order to pursue this study it is necessary to assume that the surface $\mathscr{S}$ is orientable. The reason for this restriction is made clear in the following lemma.

Index Lemma

*Let $\mathscr{S}$ be an orientable surface with boundary. Let $\mathscr{K}$ be an oriented triangulation of $\mathscr{S}$ whose vertexes are labeled A, B, and C. The **content** C of the labeling is defined as the number of complete triangles in $\mathscr{K}$; each triangle is counted plus one if the order of the vertexes ABC agrees with the orientation of the triangle in $\mathscr{K}$, and otherwise the triangle is counted minus one. The **index** I of the labeling is defined as the number of edges labeled AB on the boundary of $\mathscr{S}$; each edge is counted plus one if the order of the vertexes AB agrees with the orientation of the triangle containing that edge, and otherwise the edge is counted minus one. Then I = C.*

An example of this lemma is drawn in Figure 31.2. Note that the definition of the content and the index depend on the orientability of $\mathscr{S}$. Without orientability it would not be possible even to state the lemma. As for a proof, it is exactly the same as it was in §7 (see Exercise 1). The index lemma is needed to prove the following theorem, one of the most remarkable results in the topology of surfaces.

Poincaré Index Theorem

Let V be a continuous tangent vector field on a compact, connected, orientable surface $\mathscr{S}$. Then the sum of the indexes of the critical points of V equals the Euler characteristic of $\mathscr{S}$.

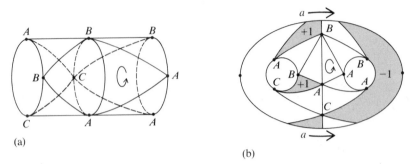

Figure 31.2   An example of the index lemma. (a) A labeled triangulation of a cylinder, $I = C = 1$. (b) The same triangulation with complete triangles shaded.

For the proof we introduce a second vector field on $\mathcal{S}$, a gradient vector field U. Figure 31.3 shows such a vector field added to the vector field of Figure 31.1. The gradient vectors U are drawn only at the critical points of V in order to avoid confusing the two vector fields. Since V has only a finite number of critical points, we can always choose the reference plane of U *not* parallel to the tangent planes at the critical points of V. Then no point can be critical for U and V simultaneously. This explains the tilted reference plane in Figure 31.3. Now, from §28, we know that the sum of the indexes of the critical points of U is the Euler characteristic of $\mathcal{S}$. The problem is to relate this sum to the corresponding sum for V.

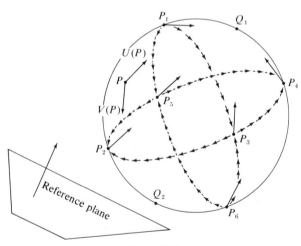

Figure 31.3

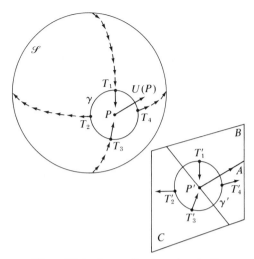

Figure 31.4 A coordinate system in the tangent plane at P.

Let us begin by considering the calculation of the index of a critical point. Figure 31.4 shows a typical critical point P for V on $\mathscr{S}$ (actually the point P_3 from Figure 31.3). By construction, $U(P)$ is *not* zero at P. To compute the index a path γ is drawn around P on the surface $\mathscr{S}$ and then projected together with the corresponding vectors V onto the tangent plane at P. In Figure 31.4 the tangent plane has been drawn at a distance from P to avoid confusion. On the surface $\mathscr{S}$ you can see the path γ, four particular points T_1, T_2, T_3, and T_4, and their vectors. The whole configuration surrounding P has then been projected onto the tangent plane where everything is labeled with primed letters. The computation of the winding number requires the establishment in the tangent plane of a direction of positive rotation and a coordinate system. The direction of positive rotation (counterclockwise in Figure 31.4) can be taken from the orientation of $\mathscr{S}$. Now the vector fields U and V are linked *by using the direction of the vector $U(P)$ to establish a coordinate system in the tangent plane, choosing $U(P)$ as the direction of the positive y-axis.* This decision determines which regions of the tangent plane are associated with each of the labels A, B, and C. In Figure 31.4 we see that the points T_1', T_2', T_3', and T_4' receive the labels $CCBA$, respectively. Accordingly, the index $I(P)$ appears to be -1, just what we expect from the phase portrait in Figure 31.1.

The next step is to dispense with the bother of projecting everything into the tangent plane. Instead we label the points T_1, T_2, T_3, and T_4 in $\mathscr{S}$

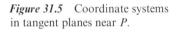

Figure 31.5   Coordinate systems
in tangent planes near *P*.

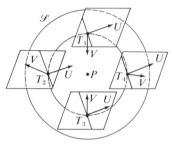

itself by establishing coordinate systems in their tangent planes. Figure 31.5 shows how this scheme applies to that critical point. Since the vector field U is continuous, at a point T near P the vector $U(T)$ will be nearly parallel to $U(P)$. Assuming that the surface $\mathscr{S}$ is smooth enough, the tangent plane at T will be nearly parallel to the tangent plane at P, and therefore the index computed by labeling points on the surface $\mathscr{S}$ will be the same as that computed by labeling points in the tangent plane at P. A technical argument, which we omit, is required to establish this point.

We can now apply the index lemma to the surface $\mathscr{S}$. According to the procedure outlined in the preceeding paragraphs, every point P on $\mathscr{S}$ is to be labeled A, B, or C depending on the direction of the vector $V(P)$ with respect to the coordinate system obtained by taking $U(P)$ as the positive y-axis and orienting the quadrants according to the orientation of $\mathscr{S}$. This labeling procedure fails only if either $V(P) = 0$ or $U(P) = 0$, that is, at the critical points of the two vector fields. Therefore let us consider the surface $\mathscr{S}'$ obtained from $\mathscr{S}$ by cutting out small disks around each of these critical points.

We now take a triangulation $\mathscr{K}$ of $\mathscr{S}'$ and apply the index lemma (see Figure 31.6). The first thing to note is that, if the triangles of $\mathscr{K}$ are sufficiently small, then there will be *no* complete triangles. Otherwise there exist arbitrarily small complete triangles on $\mathscr{S}'$, and using a familiar compactness argument (Exercise 2), this implies that $\mathscr{S}'$ contains a critical point for U or V. Since this is impossible, it follows that for sufficiently fine triangulations the content is zero. Therefore, by the index lemma, the index is also zero.

The index consists of separate contributions from each of the boundary curves of $\mathscr{S}'$, one curve around each of the critical points of V and U. Let $P_1, P_2, \ldots, P_n$ be the critical points of V. If the vertexes of $\mathscr{K}$ are choosen sufficiently close together on the paths around the points P_i, then, by our choice of labeling procedure, the path around the point P_i contributes to the index of $\mathscr{K}$ an amount equal to the index of P_i as a critical point of V. Let $Q_1, Q_2, \ldots, Q_m$ be the critical points of U. We need to find a similar interpretation for the contribution to the index of $\mathscr{K}$ of the paths around the points Q_j. We will then have determined the whole index of $\mathscr{K}$, which can be set equal to zero according to the conclusion of the preceeding paragraph.

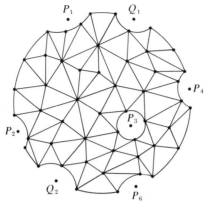

Figure 31.6

Let us examine what is happening at one of the critical points of U. Figure 31.7 displays such a point together with several examples of our labeling procedure applied to points on a path around Q. Figure 31.5 and Figure 31.7 are essentially the same, but for an interchange in the roles of U and V. Thus in Figure 31.5 we have a critical point P of V, whereas Figure 31.7 contains a critical point Q of U. In 31.5 the vector $U(P)$ is not zero, so that the vectors $U(T)$ near P are all nearly parallel; while in 31.7 the vector $V(Q)$ is not zero, so that the vectors $V(T)$ near Q are nearly parallel. In Figure 31.5 the near parallelism of the vectors U makes them suitable as reference for setting up coordinate systems in the tangent planes, leading to the computation of the index of P with respect to V. In Figure 31.7 the near parallelism of the vectors $V(T)$ would make them suitable as reference for setting up parallel coordinate systems, which would lead to the computation of the index of Q with respect to U. However, in Figure 31.7 we are *not* using

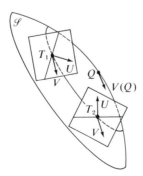

Figure 31.7

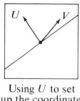

Using U to set up the coordinate system, the label is A.

Using V, the label is B.

Using U to set up the coordinate system, the label is C.

Using V, the label is C.

Figure 31.8   **Figure 31.9**

the vectors V to establish the coordinate systems, but are still using the vectors U just as in 31.5, which is why the coordinate systems in Figure 31.7 are *not* parallel. What is the effect of not using the vectors V to establish the coordinate systems near Q? Examine a few trial sketches (Figures 31.8 and 31.9) and you will be convinced of the wonderful circumstance that the only difference between the two coordinate systems possible at each point (the one using U, the other using V to determine the positive y-axis) is that vertexes labeled A under one scheme are labeled B under the other and vice versa, while vertexes labeled C will be labeled C no matter which system is used. Therefore edges labeled AB under one system are labeled BA under the other.

It follows from the preceding paragraph that the contributions to the index of the triangulation $\mathscr{K}$ of $\mathscr{S}'$ from the paths around the critical points of U is the negative of the indexes of these critical points. We have thus proven that

$$I_V(P_1) + I_V(P_2) + \cdots + I_V(P_n) - I_U(Q_1) - I_U(Q_2) - \cdots - I_U(Q_m) = 0$$

Therefore

$$I_V(P_1) + I_V(P_2) + \cdots + I_V(P_n) = I_U(Q_1) + I_U(Q_2) + \cdots + I_U(Q_m) = \chi(\mathscr{S})$$

This proves the theorem.

COMBING THE HAIRY SPHERE

On the sphere the Poincaré index theorem implies that, since the sum of the indexes of the critical points of a vector field is two, there must be at least one critical point. This is one of the most famous theorems in topology.

Corollary (Poincaré, Brouwer)

Every continuous tangent vector field on the sphere has a critical point.

It follows that it is impossible to comb the hair on a coconut. For supposing the hair on the coconut were straight, it would form a vector field, although not necessarily a field of tangents. To comb the hair would be to try to convert it to a tangent vector field—an impossibility if the hair were growing from every point of the coconut! Hence, to quote another popular phrase, the impossibility of combing the hairy sphere. A more poetic application of the theorem is this: there is always a place somewhere on the earth's surface where the wind is calm.

Exercises

1. Prove the index lemma.
2. Supply the details for the compactness argument required to complete the proof of the Poincaré index theorem.
3. For each of the following phase portraits, mark all critical points, determine their indexes, and by classifying the surface and computing its Euler characteristic, verify the conclusion of the Poincaré index theorem.

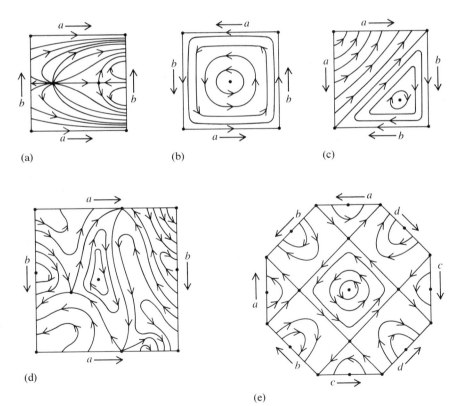

(a)   (b)   (c)

(d)

(e)

4. Draw phase portraits on the sphere to the following specifications:

(a) with just one critical point

(b) with a center and a node

5. Draw phase portraits on the torus to the following specifications:

(a) with a center and a saddle point

(b) with a dipole and two saddle points

6. Which surfaces have phase portraits without critical points? Draw some.

7. Which surfaces have phase portraits with exactly one critical point? Which have phase portraits with exactly two critical points?

8. Prove that an arbitrary vector field on the sphere either has a critical point or has a vector perpendicular to the surface of the sphere.

9. What is the total number of mountain peaks plus valley bottoms on the whole earth's surface (continental and oceanic) if we subtract the number of cross points counted by index?

Notes. Homology, as one of the fundamental branches of algebraic topology, has been the subject of many books. The most geometric of these is the classic *Lehrbuch der Topologie* by Seifert and Threlfall [27], unfortunately available only in German. More modern treatments tend to be quite algebraic. Among these, the books by Greenberg [9] and Vick [29] are excellent.

The applications presented in this chapter are the subject of several popular treatments. You should consult the references cited in Chapter One, particularly the books by Frechet and Fan [8] and Blackett [4]. Those interested in graph theory should consult Harary [10]. The book by Klein [17] deserves special mention. Here the inventor of the famous bottle applies the Poincaré index theorem to a discussion of some topics in advanced complex analysis, illustrated profusely with phase portraits on tori of one, two, and even three holes.

Ringel and Young's proof of Heawood's conjecture can be found in Ringel's book *Map Color Theorem* (Springer-Verlag, New York, 1974).

six

Continuous Transformations

§32 COVERING SPACES

After their brief treatment in Chapters One and Two, we have rather lost sight of continuous transformations. Their return is overdue considering their importance. After all, topology is defined by means of continuous transformations. Using continuous transformations it is possible to prove results of great generality. The main theorem of this chapter includes as corollaries both the Brouwer fixed point theorem and the Poincaré index theorem.

In order to appreciate the theory, it is necessary to have lots of examples. These will be provided through the use of covering spaces. Not only does the definition of covering space involve an interesting type of transformation, but covering spaces enable examples of transformations on the plane to be "set down" as transformations of other surfaces.

Definition

*Let $\mathcal{U}$ and $\mathcal{S}$ be surfaces. $\mathcal{U}$ is a **covering space** for $\mathcal{S}$ if there is a continuous transformation t from $\mathcal{U}$ onto $\mathcal{S}$ such that for each point Q of $\mathcal{S}$ there are a certain number of points P of $\mathcal{U}$ (possibly an infinite number) such that $t(P) = Q$, and each P has a neighborhood that via t is topologically equivalent to a neighborhood of Q. The transformation t is called a **covering transformation**.*

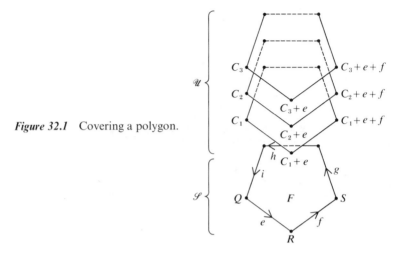

Figure 32.1   Covering a polygon.

It is easy to construct trivial examples of covering spaces for any surface by letting $\mathcal{U}$ be any number of completely separate copies of $\mathcal{S}$, called **sheets**, that can be thought of as stacked directly above $\mathcal{S}$, as in Figure 32.1. The covering transformation t takes each point of $\mathcal{U}$ to the point of $\mathcal{S}$ lying directly beneath. This picture is useful in visualizing all covering spaces. However, in interesting cases it is possible to sew the sheets of the covering space together in such a way as to obtain a connected covering space. Here is the outline of such a construction, based on the first homology group $H_1(\mathcal{S})$.

Let the connected surface $\mathcal{S}$ be divided into polygons, edges, and vertexes—in a word, presented as a complex $\mathcal{K}$. We construct the covering surface $\mathcal{U}$ as a complex, simplex by simplex, so that vertexes of $\mathcal{U}$ lie above vertexes of $\mathcal{S}$, edges lie above edges, and faces lie above faces.

THE VERTEXES OF $\mathcal{U}$

Pick a vertex P of $\mathcal{K}$ as base vertex. Every other vertex Q of $\mathcal{K}$ is connected to P by 1-chains on $\mathcal{K}$; that is, there are chains $C \in C_1(\mathcal{K})$ such that $\partial(C) = Q - P$. Some of these chains are perhaps homologous. Let all the 1-chains from P to Q be divided into homology classes: two chains are in the same class if and only if they are homologous. Then let $C_1, C_2, \ldots, C_n$ be a (possibly infinite) list of 1-chains from P to Q, one from each homology class. For each chain in this list we put a vertex in $\mathcal{U}$ over Q (see Figure 32.1), each vertex labeled by the corresponding homology class. In practice, the paths $C_1, C_2, \ldots, C_n$ may themselves be used as labels, provided it is understood that homologous paths can be substituted at any time. To summarize: for each vertex Q of $\mathscr{S}$, $\mathscr{U}$ contains n vertexes above Q, one for each homologously distinct class of paths from the base point P to Q.

THE EDGES OF $\mathscr{U}$

Let e be an edge of $\mathscr{S}$ from Q to another vertex R of $\mathscr{K}$. Suppose e is directed from Q to R (otherwise we consider $-e$). For every chain C from P to Q there is the chain $C + e$ from P to R. This establishes a correspondence between the chains from P to Q and the chains from P to R, a correspondence that preserves homology; that is, if C and D connect P and Q and $C \sim D$, then $C + e \sim D + e$. It follows that there are just as many homology classes of paths from P to Q as from P to R, hence just as many points of $\mathscr{U}$ above R as above Q. Furthermore, these paths are arranged in matched pairs C and $C + e$. For each such pair we define an edge of $\mathscr{U}$ from C to $C + e$.

THE FACES OF $\mathscr{U}$

Let F be a face of $\mathscr{S}$ containing the edge e. The vertexes of $\mathscr{U}$ lying above the vertexes of F, according to the previous paragraph, arrange themselves into matched pairs on edges of $\mathscr{U}$ lying above edges of F. These edges and vertexes are assembled into faces above F as follows: starting with a given vertex above Q, say C_1, we follow an edge above e to another vertex $C_1 + e$. Following the next edge of F(f in Figure 32.1), we obtain a vertex $C_1 + e + f$ lying above the next vertex S of F. Continuing in this way we eventually return to the vertex C_1 above Q by successively adding on the whole boundary of F, since $C_1 + e + f + g + h + i \sim C_1$. Thus the vertexes and edges of $\mathscr{U}$ lying above those of F partition themselves into disjoint cycles, one for each homology class of paths from P to Q. For each such cycle we define a face of $\mathscr{U}$ lying above F, whose boundary is the given cycle.

This completes the construction of the covering space $\mathscr{U}$. The definition of the covering transformation is obvious. Since each vertex and edge of $\mathscr{U}$ belongs to the same number of edges and faces as the vertex or edge lying beneath it in $\mathscr{S}$, it is clear that $\mathscr{U}$ is a surface. From the definition of the vertexes of $\mathscr{U}$, we see that if two vertexes of $\mathscr{S}$ lie on an edge together, then they have the same number of vertexes of $\mathscr{U}$ lying above them. Since $\mathscr{S}$ is connected, any two vertexes of $\mathscr{S}$ can be connected by a path of edges on $\mathscr{K}$. Therefore *any* two vertexes have the same number of vertexes of $\mathscr{U}$ lying above them. It follows, then, from the definition of the edges and faces of $\mathscr{U}$ that every point of $\mathscr{S}$ is covered by the same number of points of $\mathscr{U}$. Thus $\mathscr{U}$ is a covering space for $\mathscr{S}$.

The number of points of $\mathcal{U}$ covering a given point of $\mathcal{S}$ is best determined by counting the number of points of $\mathcal{U}$ lying above the base point P. Vertexes of $\mathcal{U}$ lying above P correspond to the homology classes of chains from P to P, that is, to homology classes of cycles. Therefore each point of $\mathcal{S}$ is covered by as many points as there are elements in the first homology group $H_1(\mathcal{S})$.

In order to prove that $\mathcal{U}$ is connected, it suffices to prove that all the vertexes of $\mathcal{U}$ can be connected to some one vertex. Let Φ be the vertex of $\mathcal{U}$ lying above the base point P and corresponding to the null cycle $\varnothing$. Let C be any other vertex of $\mathcal{U}$. Then C is also the name for a 1-chain on $\mathcal{K}$ from P to some other vertex Q of $\mathcal{K}$. The chain C can be written as a sum of edges $C = e_1 + e_2 + \cdots + e_k$, where e_1 connects P to some vertex P_1, e_2 connects P_1 and P_2, and so on until e_k connects P_{k-1} and $P_k = Q$. The sequence of partial sums $C_0 = \varnothing$, $C_1 = e_1$, $C_2 = e_1 + e_2, \ldots, C_k = C = e_1 + e_2 + \cdots + e_k$ represents a sequence of vertexes of $\mathcal{U}$ starting with $C_0 = \varnothing = \Phi$ and ending with C, in which adjacent vertexes, say C_j and C_{j+1}, are connected by an edge lying above e_{j+1}. In other words, the sequence $C_0, C_1, \ldots, C_k$ is the sequence of vertexes of a chain in $\mathcal{U}$ connecting P and C. This proves $\mathcal{U}$ is connected. The covering space $\mathcal{U}$ has many remarkable properties. It is called the **universal covering surface** of $\mathcal{S}$.

Examples

1. *The projective plane.* The first homology group of the projective plane has just two elements; therefore the universal covering $\mathcal{U}$ of the projective plane is a two-sheeted covering. Let $\mathcal{P}$ be the projective plane (Figure 32.2). $\mathcal{P}$ has one vertex, one edge, and one face. Therefore $\mathcal{U}$ has two vertexes, two edges, and two faces. The two elements of $H_1(\mathcal{P})$ are the null chain $\varnothing$ and the single loop a. Let the two vertexes of $\mathcal{U}$ above the single vertex P of $\mathcal{P}$ be labeled Φ and A. The one edge of $\mathcal{P}$ appears twice; therefore the two edges of $\mathcal{U}$ appear twice, once over each appearance of the edge a in $\mathcal{P}$. Let c be the edge of $\mathcal{U}$ from Φ to $\Phi + A = A$, and let b be the edge of $\mathcal{U}$ from A to $A + A = \Phi$. It follows that each face of $\mathcal{U}$ contains both edges b and c with the identifications shown in Figure 32.2 Surprisingly enough, these identifications can actually be carried out in space (Figure 32.3b) if the upper face is rotated 180° (Figure 32.3a). It turns out that $\mathcal{U}$ is a sphere! The two faces of $\mathcal{U}$ are two hemispheres. Points on the upper hemisphere no longer lie directly above the points they are intended to cover. Instead the two points of $\mathcal{U}$ covering a point of $\mathcal{P}$ now are diametrically opposite each other. Thus the projective plane can be regarded as obtained from a sphere by sewing together diametrically opposed points.

2. *The torus.* Using the plane model $\mathcal{T}$ of the torus, we have one vertex, two edges, and one face. The universal covering surface $\mathcal{U}$ will have an infinite number of vertexes, one for each of the cycles $ma + nb$, where a and

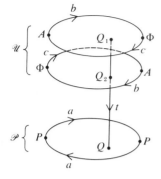

Figure 32.2   Two-sheeted covering of the projective plane.

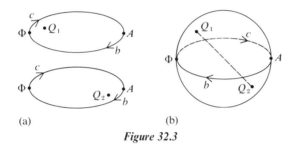

(a)   (b)

Figure 32.3

b are the two edges of $\mathcal{T}$, and m, n are any integers (positive, negative, or zero). The vertexes of $\mathcal{U}$ can thus be laid out on a rectangular grid in a plane (Figure 32.4). Every horizontal (or vertical) line between grid points corresponds to adding or subtracting a (or b) and hence corresponds to an edge of $\mathcal{U}$. The faces of $\mathcal{U}$ come out as congruent rectangles covering the whole plane; thus $\mathcal{U}$ is a plane.

3. *The double torus.* The universal covering spaces of other surfaces are more difficult to describe. Of course, the more simplexes in the original surface and the larger the first homology group $H_1(\mathcal{S})$, the more complicated the structure of $\mathcal{U}$. Figure 32.5 shows how octagons may be sewn together to form the universal covering surface of the double torus. There are an infinite number of octagons. They fill out the open disk shown in Figure 32.5 just as the rectangles in Figure 32.4 fill out the plane. Since the plane and the open disk are topologically equivalent (§2), the double torus and the torus have the same universal covering space, although the internal construction of these spaces is very different.

In a similar fashion, the plane turns out to be the universal covering space for all the connected sums of tori. Since the plane with a point added at

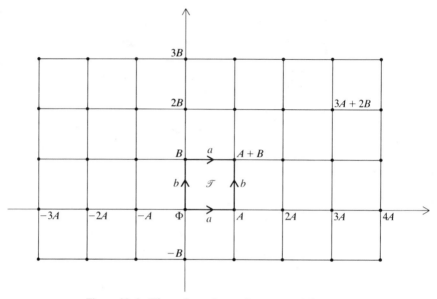

Figure 32.4 The universal covering space of the torus.

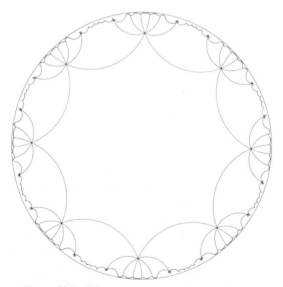

Figure 32.5 The universal covering space of the double torus.

infinity also covers the sphere, *the plane covers all the orientable surfaces*. In addition, the plane covers the projective plane, since the sphere covers the projective plane. Thus covering spaces point to the plane as *the* surface the study of which is crucial to the understanding of other surfaces. This is something we knew already, but it is interesting to have it confirmed in this dramatic way.

Covering spaces can be used systematically to study examples of transformations on other surfaces. The surface $\mathscr{S}$ can be imbedded as one sheet of its universal covering space $\mathscr{U}$, then the covering space can be subjected to a continuous transformation f, and the whole thing projected back to $\mathscr{S}$ by the covering transformation t. The result of this composition is a transformation of the original surface $\mathscr{S}$. In this way a correspondence is set up between certain transformations of $\mathscr{S}$ and certain transformations of the covering space $\mathscr{U}$. We shall use this device to produce interesting examples of transformations. The only restriction is that the transformation f of the covering space must have this property: if P and Q cover the same point of $\mathscr{S}$, then $f(P)$ and $f(Q)$ must also cover the same point of $\mathscr{S}$. Figure 32.6 gives an example. The torus is first imbedded as a rectangle in the covering space (Figure 32.6b), then subjected to a plane transformation (a translation by the vector A, Figure 32.6c). The translation has the property that two points covering the same point of $\mathscr{T}$ (say P and P') continue to cover the same point (Q) after translation. After returning to $\mathscr{T}$ by the covering transformation, we obtain a transformation r_A of the torus, called a **rotation** of the torus.

The covering space can be used to discuss other aspects of a surface $\mathscr{S}$. For example, a phase portrait on the projective plane (Figure 32.7a) can be

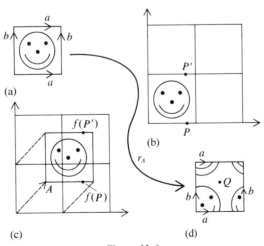

Figure 32.6

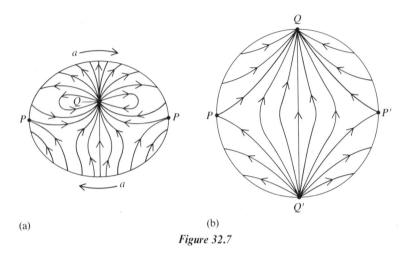

(a)   (b)

Figure 32.7

lifted to a phase portrait on the covering sphere (Figure 32.7b). The resulting phase portrait on the sphere is central symmetric, meaning that it has exactly the same appearance at diametrically opposite points of the sphere. For example, every critical point from the original phase portrait on $\mathscr{P}$ appears twice on the sphere $\mathscr{U}$. Since the sum of the indexes of these critical points on the sphere is two, it follows that the sum of the indexes of the critical points on $\mathscr{P}$ must have been one. This extends to the projective plane the conclusion of the Poincaré index theorem.

Theorem

The sum of the indexes of the critical points of a phase portrait on the projective plane equals the Euler characteristic of the projective plane.

Exercises

1. Find the universal covering spaces for the sphere, disk, cylinder, Klein bottle, and Möbius strip.

2. Using homology (mod 2) instead of integral homology, find another covering space for the torus. Can you find still more covering spaces for the torus?

3. Using the model of the torus in space, describe the rotation r_A in the special cases $A = (0, 1)$ and $A = (1, 0)$. These examples explain why r_A is called a rotation.

4. Explain how a rotation of the sphere can induce a continuous transformation of the projective plane.

5. A phase portrait in the plane is **doubly periodic** when it repeats the same pattern of curves in a rectangular grid. For the two examples shown below, shade in one of the basic rectangles whose pattern is being repeated across the plane. Explain why the sum of the indexes of the critical points in such a rectangle must be zero, and verify this for these examples.

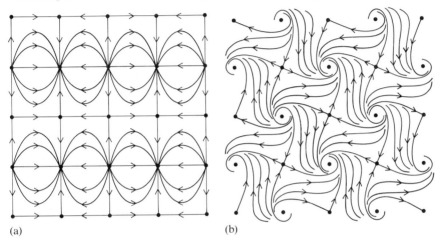

(a)   (b)

6. Draw the field lines for the following infinite arrays of charges. Verify in each case that the theorem proved in the previous exercise is satisfied. Doubly periodic electrical fields such as these arise in practice when charges are placed in a rectangular box with reflecting sides, and in studying the electrical properties of crystals.

(a)   (b)

EXAMPLES FROM COMPLEX ANALYSIS

The complex numbers $z = x + iy$ correspond to the points of the plane. Algebraic operations with the complex numbers correspond to geometric transformations of the plane. Here is another example of the interrelationships that can develop between different branches of mathematics. As with situations of a similar sort, this connection between complex analysis and topology can be exploited in two ways. On the one hand, topology can be used to prove theorems about the complex numbers, as in §37 we shall prove the fundamental theorem of algebra. On the other hand, complex algebra can be used to give examples of topological transformations on the plane and on the many other surfaces that the plane covers.

Let us begin by reviewing the basic properties of complex numbers. Complex numbers can be expressed in two convenient forms, both of which are important. The formula $z = x + iy$ is the **rectangular form** of the complex number z, since it is based on the rectangular coordinates x and y, where the y-axis is measured in units of i. The other form is $z = r(\cos \theta + i \sin \theta)$, where $r = (x^2 + y^2)^{1/2}$ is the distance from z to the origin, also called the **modulus** of z and denoted $\|z\|$, and θ is the **vectorial angle** (see Figure 32.8). This is called the **polar form** of z, since it is based on the polar coordinates r and θ. Introducing the **complex exponential**, $e^{i\theta} = \cos \theta + i \sin \theta$, the polar form becomes $z = \|z\| e^{i\theta}$.

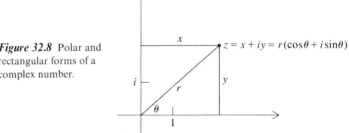

Figure 32.8 Polar and rectangular forms of a complex number.

Addition of complex numbers amounts to no more than familiar vector addition: if $z_1 = x_1 + iy_1$ and $z_2 = x_2 + iy_2$, then $z_1 + z_2 = (x_1 + x_2) + i(y_1 + y_2)$ (see Figure 32.9). Since the sum of two sides of a triangle is always greater than the third side, we obtain the triangle law,

$$\|z_1 + z_2\| \le \|z_1\| + \|z_2\|$$

an important property of the modulus.

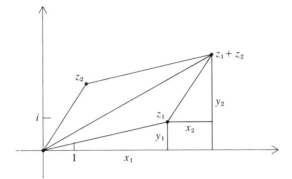

Figure 32.9 The geometry of complex addition.

Multiplication is introduced by setting $z_1 z_2 = (x_1 x_2 - y_1 y_2) + i(x_1 y_2 + x_2 y_1)$. From the equation

$$z_1 z_2 = \|z_1\| e^{i\theta_1} \|z_2\| e^{i\theta_2}$$
$$= \|z_1\| \|z_2\| e^{i(\theta_1 + \theta_2)}$$

we deduce that $\|z_1 z_2\| = \|z_1\| \|z_2\|$ and that the vectorial angle of $z_1 z_2$ is $\theta_1 + \theta_2$. This depends on the essential uniqueness of the polar form plus the familiar exponential law $e^a e^b = e^{a+b}$, which is easily proven in this case using familiar trigonometric identities. Thus we arrive at the following geometric picture of complex multiplication: the product $z_1 z_2$ is found by adding the vectorial angles of z_1 and z_2 and going out on the resulting radius vector a distance $\|z_1\| \|z_2\|$ (see Figure 32.10).

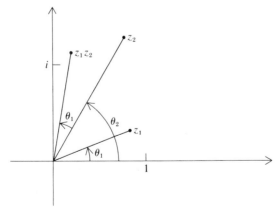

Figure 32.10 The geometry of complex multiplication.

Keeping the geometry of complex multiplication in mind, consider a complex function such as $p(z) = z^2$. p transforms z into the complex number with twice the vectorial angle of z and a modulus equal to the square of the modulus of z. In order to get a feeling for this transformation, imagine the point z traveling around the origin once clockwise on a circle of radius R. Then the image $p(z)$ will travel *twice* around the origin on a circle of radius R^2. The image of the interior of this circle of radius R under the transformation p will cover the interior of the circle of radius R^2 two times. Complex analysis thus gives rise to some interesting transformations.

Exercises

7. Prove that $e^{i\theta_1} e^{i\theta_2} = e^{i(\theta_1 + \theta_2)}$.

8. Prove directly that $\|z_1 z_2\| = \|z_1\| \, \|z_2\|$.

9. To what familiar geometric transformations (such as rotations, translations, reflections, etc.) do these complex functions correspond?

(a) $p(z) = z + 1$ (b) $p(z) = -z$

(c) $p(z) = \bar{z}$ (d) $p(z) = iz$

(e) $p(z) = 2z$ (f) $p(z) = -\bar{z}$

(g) $p(z) = z + i$ (h) $p(z) = e^{i\theta} z$

The quantity $\bar{z} = x - iy$ is called the **complex conjugate** of z.

10. Let $p(z) = a_0 + a_1 z + \cdots + a_n z^n$ be a polynomial on the complex plane. Define $p(\infty) = \infty$. Explain how p induces a continuous transformation on the sphere. Let $q(z) = b_0 + b_1 z + \cdots + b_m z^m$ be a second polynomial, and let $f(z)$ be the rational function $f(z) = p(z)/q(z)$. How should $f(\infty)$ be defined in order to obtain a continuous transformation of the sphere?

11. The rational function

$$f(z) = \frac{az + b}{cz + d}$$

where a, b, c, and d are complex numbers such that $ad - bc \neq 0$, is called a **Möbius transformation**. Show that f is a one-to-one continuous transformation onto the sphere. Show that the inverse transformation is also a Möbius transformation. Therefore deduce that f is a topological transformation of the sphere.

12. Show that in the stereographic projection of the sphere onto the plane, the endpoints of a diameter of the sphere correspond to points z and z^* in the plane such that $\bar{z} z^* = -1$.

13. Show that the complex plane with the point at infinity added is a double covering for the projective plane in which the points z and z^* cover the same point on the projective plane if $\bar{z}z^* = -1$.

14. Let f be a continuous transformation of the complex plane satisfying the identity

$$f(-1/\bar{z}) = -1/\overline{f(z)} \tag{1}$$

Prove that if $\bar{z}z^* = -1$, then also $\overline{f(z)}f(z^*) = -1$. Explain how f induces a continuous transformation of the projective plane.

15. Show that the Möbius transformation

$$f(z) = \frac{az + b}{-\bar{b}z + \bar{a}}$$

satisfies (1). When $\|a\|^2 + \|b\|^2 = 1$, this transformation actually represents a rotation of the sphere transferred by stereographic projection to the complex plane. In this case find the endpoints in the plane of the diameter around which this rotation takes place. The transformation induced by f in the projective plane is called a **rotation** of the projective plane.

16. Show that the functions $f(z) = z^3, z^5, \ldots, z^{2n+1}$ all satisfy (1). Consider also $f(z) = 1/z^3, 1/z^5, \ldots, 1/z^{2n+1}$. The identity $\overline{zw} = \bar{z} \cdot \bar{w}$ will come in handy. Prove it first.

17. Show that the composition of functions satisfying (1) also satisfies (1).

18. A complex function f is called **doubly periodic** if there exist real numbers a and b such that $f(z + a) = f(z)$ and $f(z + ib) = f(z)$. Explain how such a function induces a transformation from the torus to the complex plane. Show that a doubly periodic function has no limit at infinity. Deduce that no polynomial or rational function can be doubly periodic.

---

§33 SIMPLICIAL TRANSFORMATIONS

The object of this chapter is to conduct a general study of continuous transformations, at least to the extent that they can be attacked using homology. In the process, proofs will become more and more algebraical, reflecting the transition from **combinatorial topology** to **algebraic topology** that began in the decade 1925–1935 with the introduction of group theory into topology.

We begin by studying simplicial transformations. These transformations are especially suited to homology and form a connecting link between continuous transformations of surfaces on the one hand and group homo-morphisms of the homology groups on the other. A simplicial transformation is a transformation from one complex $\mathscr{K}$ to another $\mathscr{L}$ that sends vertexes of $\mathscr{K}$ to vertexes of $\mathscr{L}$, edges to edges, and faces to faces. In a word, a simplicial transformation maps simplexes of $\mathscr{K}$ to simplexes of $\mathscr{L}$. In order for this to make sense, all the faces of $\mathscr{K}$ and $\mathscr{L}$ must have the same number of edges; therefore we work exclusively with triangulations. We begin by setting up a special notation for edges and faces in a triangulation.

Definitions

Let $\mathscr{K}$ be a triangulation, and let P and Q be the endpoints of an edge. Since in a triangulation there can be at most one edge between two vertexes, let the symbol $\overline{PQ}$ stand for the edge directed from P to Q, so that $\partial(\overline{PQ}) = Q - P$. Similarly, $\partial(\overline{QP}) = P - Q$. This notation eliminates the need for separate labels for the edges. In case $P = Q$, let $\overline{PQ}$ be defined as equal to $\varnothing$. In this case the equation $\partial(\overline{PQ}) = Q - P$ still holds. If P and Q do not lie on an edge together, we leave the symbol $\overline{PQ}$ undefined.

Let PQR be the three vertexes of a triangle of $\mathscr{K}$. Since in a triangulation there can be at most one triangle with three given vertexes, let the symbol $\overline{PQR}$ stand for this triangle, oriented so that $\partial(\overline{PQR}) = \overline{PQ} + \overline{QR} + \overline{RP}$. This notation eliminates the need for separate labels for the faces in a triangulation. In case two or more of the vertexes P, Q, and R are equal, while still lying together on an edge, let $\overline{PQR}$ be defined as equal to $\varnothing$, so that the equation $\partial(PQR) = \overline{PQ} + \overline{QR} + \overline{RP}$ still holds. If P, Q, and R do not lie on the same triangle together, we leave the symbol $\overline{PQR}$ undefined.

The way a simplicial transformation maps the edges and faces of a tri-angulation is determined by the way it maps the vertexes alone. This leads to the following formal definition

Definition

Let $\mathscr{K}$ and $\mathscr{L}$ be triangulations of topological spaces. A **simplicial trans-formation** τ is a function from the vertexes of $\mathscr{K}$ to the vertexes of $\mathscr{L}$ such that if P and Q are vertexes of $\mathscr{K}$ lying on a single edge of $\mathscr{K}$, then $\tau(P)$ and $\tau(Q)$ lie on a single edge of $\mathscr{L}$; and if P, Q, R are vertexes of $\mathscr{K}$ lying on a single triangle of $\mathscr{K}$, then $\tau(P)$, $\tau(Q)$, and $\tau(R)$ lie on a single triangle of $\mathscr{L}$.

Let τ be a simplicial transformation from $\mathscr{K}$ to $\mathscr{L}$. Let P and Q be vertexes of $\mathscr{K}$. If $\overline{PQ}$ is defined on $\mathscr{K}$, then the definition of simplicial transformation

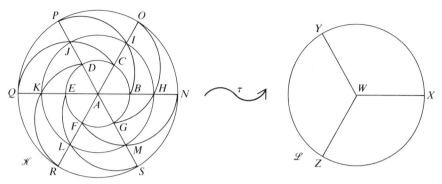

Figure 33.1

ensures that $\overline{\tau(P)\tau(Q)}$ is defined on $\mathcal{L}$. In this way τ induces a mapping τ_1 from the edges of $\mathcal{K}$ to the edges of $\mathcal{L}$ defined by the formula

$$\tau_1(\overline{PQ}) = \overline{\tau(P)\tau(Q)}$$

Similarly, τ induces a mapping τ_2 from the triangles of $\mathcal{K}$ to the triangle of $\mathcal{L}$ defined by the formula

$$\tau_2(\overline{PQR}) = \overline{\tau(P)\tau(Q)\tau(R)}$$

An example will help clarify all these notions and notations. Figure 33.1 shows two triangulations of a disk. In the triangulation $\mathcal{K}$, the edges $\overline{AB}$, $\overline{AC}$ issue outward from the center, while $\overline{PQ}$, $\overline{QR}$ are edges on the boundary directed counterclockwise. The triangle $\overline{ABC}$ lies near the center of the disk and is also directed counterclockwise. The symbols $\overline{PDQ}$, $\overline{ES}$, and $\overline{DAT}$ are all undefined. The table below gives a simplicial transformation from $\mathcal{K}$ to $\mathcal{L}$:

	A	B	C	D	E	F	G	H	I	J	K	L	M	N	O	P	Q	R	S
τ	W	W	W	W	W	W	W	X	Y	Z	X	Y	Z	X	Y	Z	X	Y	Z

Some examples of the action of the induced transformations are $\tau_1(\overline{AB}) = \overline{WW} = \emptyset$, $\tau_1(\overline{BH}) = \overline{WX}$ and $\tau_2(\overline{ABC}) = \emptyset$, $\tau_2(\overline{BHI}) = \overline{WXY}$, $\tau_2(\overline{BHG}) = \emptyset$. You can easily add further examples and complete a table showing how τ transforms all the edges and triangles of $\mathcal{K}$.

The mappings of vertexes, edges, and faces set up by a simplicial transformation extend by additivity to give three homomorphisms τ_k ($k = 0, 1, 2$) of the corresponding chain groups, $\tau_k : C_k(\mathcal{K}) \to C_k(\mathcal{L})$. Thus, for example, referring to Figure 33.1,

$$\tau_0(A + B + H) = 2W + X$$
$$\tau_1(\overline{RQ} - 2\overline{NO}) = 3\overline{YX}$$
$$\tau_2(\overline{PQJ} - \overline{QJK}) = \varnothing$$

Concerning these homomorphisms, the important thing is to establish their relationship with the boundary operator. This is given by the equations

$$\partial\tau_1 = \tau_0\partial \tag{1}$$
$$\partial\tau_2 = \tau_1\partial \tag{2}$$

based on the following diagram:

$$
\begin{array}{ccccc}
C_2(\mathcal{K}) & \xrightarrow{\partial} & C_1(\mathcal{K}) & \xrightarrow{\partial} & C_0(\mathcal{K}) \\
\downarrow{\scriptstyle \tau_2} & & \downarrow{\scriptstyle \tau_1} & & \downarrow{\scriptstyle \tau_0} \\
C_2(\mathcal{L}) & \xrightarrow{\partial} & C_1(\mathcal{L}) & \xrightarrow{\partial} & C_0(\mathcal{L})
\end{array}
$$

The composition $\partial\tau_1$, for example, corresponds to the path from $C_1(\mathcal{K})$ down to $C_1(\mathcal{L})$, then across to $C_0(\mathcal{L})$, while the composition $\tau_0\partial$ corresponds to the path across to $C_0(\mathcal{K})$ and then down to $C_0(\mathcal{L})$. Equation (1) thus says that the homomorphisms obtained going by these two paths from $C_1(\mathcal{K})$ to $C_0(\mathcal{L})$ are equal. Similarly, (2) asserts that the homomorphisms corresponding to the two paths from $C_2(\mathcal{K})$ to $C_1(\mathcal{L})$ are equal. The situation is described by saying: the diagram **commutes**. To prove (1), it suffices by additivity to verify the equation for a simplex $\overline{PQ}$ alone:

$$\partial\tau_1(\overline{PQ}) = \partial(\overline{\tau(P)\tau(Q)}) = \tau(Q) - \tau(P)$$
$$\tau_0\partial(\overline{PQ}) = \tau_0(Q - P) = \tau(Q) - \tau(P)$$

Equation (2) has a similar proof.

It follows from (1) and (2) that each transformation τ_k maps boundaries to boundaries and cycles to cycles. Consider τ_1, for example. Let c be a 1-boundary. Then $c = \partial(B)$ for some 2-chain B. Using (2) we find that $\tau_1(c) = \tau_1(\partial(B)) = \partial(\tau_2(B))$, therefore $\tau_1(c)$ is the boundary of $\tau_2(B)$. Now let c be a 1-cycle. Then $\partial\tau_1(C) = \tau_0\partial(c) = \varnothing$ so that $\tau_1(c)$ is also a cycle.

It follows next that each homomorphism preserves the homology relation; that is, if c_1 is homologous to c_2 on $\mathcal{K}$, then $\tau_k(c_1)$ is homologous to $\tau_k(c_2)$

on $\mathscr{L}$. Thus if c_1 and c_2 are cycles, then $\tau_k(c_1)$ and $\tau_k(c_2)$ are also cycles, and if $c_1 - c_2$ is a boundary, then $\tau_k(c_1) - \tau_k(c_2) = \tau_k(c_1 - c_2)$ is also a boundary. This proves the following theorem.

Theorem 1

A simplicial transformation τ induces group homomorphisms τ_k from $H_k(\mathscr{K})$ to $H_k(\mathscr{L})$.

Exercises

1. Let P be a point in a space $\mathscr{S}$ with triangulation $\mathscr{K}$. The **star of P in $\mathscr{K}$**, written st(P), is defined to be the set of points Q such that every simplex containing Q contains P. Draw examples of the three types of stars, depending on whether the point P is a vertex of $\mathscr{K}$, is not a vertex but lies on an edge, or lies in the interior of a triangle of $\mathscr{K}$. Show that st(P) is always an open set containing P. For two points P, Q show that st(P) $\cap$ st(Q) $\neq \emptyset$ if and only if P and Q lie together on a triangle of $\mathscr{K}$.

2. Let τ be a map from the vertexes of a triangulation $\mathscr{K}$ to the vertexes of a second triangulation $\mathscr{L}$. Prove that if τ is a simplicial transformation, then for any three vertexes P, Q, R of $\mathscr{K}$,

$$\mathrm{st}(P) \cap \mathrm{st}(Q) \cap \mathrm{st}(R) \neq \emptyset$$

implies

$$\mathrm{st}(\tau(P)) \cap \mathrm{st}(\tau(Q)) \cap \mathrm{st}(\tau(R)) \neq \emptyset$$

Conversely, if τ has this property, prove that τ is simplicial.

3. Prove equation (2), and show that τ_0 and τ_2 preserve boundaries and cycles.

SIMPLICIAL APPROXIMATION

The homology groups provide an algebraic picture of a topological space $\mathscr{S}$, a sort of group photograph from which information about the topological space may be deduced. Given two topological spaces $\mathscr{S}$ and $\mathscr{T}$ and a continuous transformation $f : \mathscr{S} \to \mathscr{T}$, we shall define (at least for triangulable spaces) group homomorphisms $H_k(f)$ from $H_k(\mathscr{S})$ to $H_k(\mathscr{T})$ ($k = 0, 1, 2$). Just as the homology groups $H_k(\mathscr{S})$ provide an algebraic picture of the space

$\mathscr{S}$, so the group homomorphisms $H_k(f)$ provide an algebraic picture of the transformation f.

Simplicial transformations are the key to defining the group homomorphisms $H_k(f)$. According to Theorem 1, each simplicial transformation *does* lead to group homomorphisms of the homology groups. The problem is to find simplicial transformations that are in some sense close to a given continuous transformation f. These simplicial approximations of f can then be used to define $H_k(f)$.

Definition

*Let f be a continuous transformation from a topological space $\mathscr{S}$ to a space $\mathscr{T}$. Let $\mathscr{K}$ and $\mathscr{L}$ be triangulations of $\mathscr{S}$ and $\mathscr{T}$, respectively. A **simplicial approximation of f** is a simplicial transformation τ from $\mathscr{K}$ to $\mathscr{L}$ such that for each vertex P of $\mathscr{K}$*

$$f(\mathrm{st}(P)) \subseteq \mathrm{st}(\tau(P))$$

This condition ensures not only that $f(P)$ will be near $\tau(P)$ but also that f takes a whole neighborhood of P (i.e., $\mathrm{st}(P)$) into a neighborhood of $\tau(P)$ (i.e., $\mathrm{st}(\tau(P))$).

As an example of simplicial approximation, consider the transformation $f(z) = z^2$ of the unit disk in the complex plane. The simplicial transformation τ of Figure 33.1 is a simplicial approximation of f. Figure 33.2 shows how f actually transforms the vertexes of $\mathscr{K}$ for the purpose of comparison with τ.

You can now see in outline the basic line of attack in this chapter. Given a continuous transformation $f : \mathscr{S} \to \mathscr{T}$, we choose triangulations $\mathscr{K}$ and $\mathscr{L}$ and a simplicial approximation $\tau : \mathscr{K} \to \mathscr{L}$ for f. The approximation τ induces group homomorphisms $H_k(\mathscr{S}) \to H_k(\mathscr{T})$ that are actually determined by f and hence will be denoted $H_k(f)$. Then we will try to investigate properties of the original continuous transformation f by studying algebraic properties of the homomorphisms $H_k(f)$. A necessary preliminary is to establish the existence of simplicial approximations for f. This is done in the following theorem.

Theorem 2

Let f be a continuous transformation from the compact space $\mathscr{S}$ to the space $\mathscr{T}$. Let $\mathscr{K}$ and $\mathscr{L}$ be triangulations of these spaces. Then there is a subdivision $\mathscr{K}^+$ of $\mathscr{K}$ on which there is a simplicial approximation $\tau : \mathscr{K}^+ \to \mathscr{L}$ for f.

The proof depends on the existence of a subdivision $\mathscr{K}^+$ of $\mathscr{K}$ with the following property: (*) for each vertex P of $\mathscr{K}^+$, $f(\mathrm{st}(P))$ is entirely contained

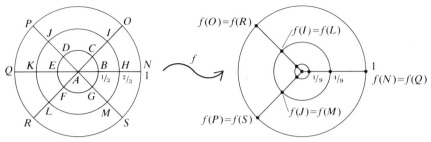

Figure 33.2

in st(Q) for at least one vertex Q of $\mathscr{L}$. Suppose for the moment that a subdivision $\mathscr{K}^+$ with property (*) exists. Then a simplicial approximation for f is defined by setting $\tau(P) = Q$ for each vertex $P \in \mathscr{K}^+$, where Q is any one of the vertexes of $\mathscr{L}$ whose existence is guaranteed by property (*). If τ is so defined, then property (*) becomes (**) $f(\mathrm{st}(P)) \subseteq \mathrm{st}(\tau(P))$ for each vertex P of $\mathscr{K}^+$. Thus τ is certainly a simplicial approximation for f, as soon as we check that τ *is* a simplicial transformation. To do this, let P, Q, R be three vertexes of $\mathscr{K}^+$ such that $\mathrm{st}(P) \cap \mathrm{st}(Q) \cap \mathrm{st}(R) \neq \varnothing$. Then

$$\mathrm{st}(\tau(P)) \cap \mathrm{st}(\tau(Q)) \cap \mathrm{st}(\tau(R)) \supseteq f(\mathrm{st}(P)) \cap f(\mathrm{st}(Q)) \cap f(\mathrm{st}(R))$$
$$\supseteq f(\mathrm{st}(P) \cap \mathrm{st}(Q) \cap \mathrm{st}(R)) \neq \varnothing$$

According to Exercise 2, this proves that τ is simplicial; hence τ is a simplicial approximation to f.

It remains to prove that a subdivision $\mathscr{K}^+$ with property (*) exists. This depends on the following compactness argument. Consider the sequence of triangulations $\mathscr{K}_0, \mathscr{K}_1, \mathscr{K}_2, \ldots$, where $\mathscr{K}_0 = \mathscr{K}$ and each $\mathscr{K}_n$ is the barycentric subdivision of $\mathscr{K}_{n-1}$ (see Figure 22.1). The sides of the triangles of $\mathscr{K}_n$ are thus at most half the length of the sides of the preceeding triangulation $\mathscr{K}_{n-1}$. If none of these triangulations has property (*), then every triangulation $\mathscr{K}_n$ contains a vertex P_n such that $f(\mathrm{st}(P_n))$ is not contained in any set $\mathrm{st}(Q)$, where Q is a vertex of $\mathscr{L}$. We will deduce a contradiction, thus proving that at least one triangulation $\mathscr{K}_n$ has property (*). By compactness, there is a point P of $\mathscr{S}$ near the set $\{P_n\}$. By continuity, $f(P)$ is near the set $\{f(P_n)\}$. Now $f(P)$, being a point of $\mathscr{T}$, must belong to $\mathrm{st}(Q)$ for some vertex Q of $\mathscr{L}$. Let R_n be a point of $\mathrm{st}(P_n)$ such that $f(R_n)$ is not in $\mathrm{st}(Q)$. The existence of such a point is guaranteed because $f(\mathrm{st}(P_n))$ is not contained in $\mathrm{st}(Q)$. Since the triangles in the triangulations $\mathscr{K}_n$ grow finer and finer, it is easy to see that P is near $\{R_n\}$. Therefore $f(P)$ is near $\{f(R_n)\}$. This contradicts the fact that $f(P)$ is in $\mathrm{st}(Q)$, which is an open set according to Exercise 1, since no points of $\{f(R_n)\}$ are in $\mathrm{st}(Q)$.

Exercises

4. Give details to support the proof of Theorem 2.

5. Supposing that a certain triangulation $\mathcal{K}^*$ has the property (*) described in the proof of Theorem 2. Show that every subdivision of $\mathcal{K}^*$ also has property (*). Thus there is a simplicial approximation to f defined on every subdivision of $\mathcal{K}^*$.

§34 INVARIANCE AGAIN

The purpose of this section is to prove the following theorem.

Invariance Theorem

*Let $f:\mathcal{S} \to \mathcal{T}$ be a continuous transformation, and let $\tau:\mathcal{K} \to \mathcal{L}$ be a simplicial approximation for f. Then the induced transformation $\tau_k:H_k(\mathcal{S}) \to H_k(\mathcal{T})$ is the same for all simplicial approximations of f, hence we say this homomorphism is **induced** by f and define $H_k(f) = \tau_k$.*

The importance of this theorem is clear. Without invariance there is no point investigating all those dull simplicial transformations; with invariance we can be certain that the homomorphisms induced are independent of the choice of simplicial transformation, that they depend only on the original continuous transformation and hence can be expected to yield information about the continuous transformation f.

Our proof of this theorem will rely on the Hauptvermutung for two-dimensional spaces. It begins with a series of lemmas, several of which prove important properties of the homology transformations. First we consider the case of two simplicial approximations to a transformation, both defined on the same triangulation.

Lemma 1

Let τ and σ be two simplicial approximations to f from the same triangulation $\mathcal{K}$ of $\mathcal{S}$ to the same triangulation $\mathcal{L}$ of $\mathcal{T}$. Then τ and σ are homologous: $\tau_k \sim \sigma_k$ $(k=0, 1, 2)$; hence τ and σ induce the same homomorphisms of the homology groups.

The proof involves the following definition and lemma.

Definition

Two simplicial transformations $\sigma, \tau: \mathcal{K} \to \mathcal{L}$ are **contiguous** *if (a) for each vertex P of $\mathcal{K}$, $\tau(P)$ and $\sigma(P)$ lie on the same edge of $\mathcal{L}$ and (b) for each edge $\overline{PQ}$ of $\mathcal{K}$, $\overline{\tau(P)\tau(Q)}$ and $\overline{\sigma(P)\sigma(Q)}$ lie on the same triangle of $\mathcal{L}$.*

Lemma 2

Contiguous simplicial transformations induce the same group homomorphisms.

With two simplicial transformations, we have the following diagram of induced transformations on the chain groups:

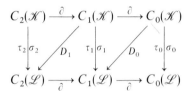

The proof involves introducing two auxiliary homomorphisms D_0 and D_1 represented above by the diagonal arrows. The resulting diagram does *not* commute. Rather we shall eventually verify the following equations:

$$\tau_0 - \sigma_0 = \partial D_0 \tag{1}$$

$$\tau_1 - \sigma_1 = \partial D_1 + D_0\partial \tag{2}$$

$$\tau_2 - \sigma_2 = D_1\partial \tag{3}$$

The homomorphism D_0 is defined by setting

$$D_0(P) = \sigma(P)\tau(P)$$

for a vertex P of $\mathcal{K}$, since the hypothesis of the lemma guarantees that $\overline{\tau(P)\sigma(P)}$ is defined, and then extending by additivity to a homomorphism $D_0: C_0(\mathcal{K}) \to C_1(\mathcal{L})$. The homomorphism D_1 is defined by setting

$$D_1(PQ) = \overline{\tau(P)\tau(Q)\sigma(Q)} - \overline{\tau(P)\sigma(P)\sigma(Q)}$$

and then extending by additivity to a homomorphism $D_1: C_1(\mathcal{K}) \to C_2(\mathcal{L})$.

In order to prove equations (1), (2), and (3), it suffices to verify each for a single simplex. For example we see that

$$\partial D_0(P) = \partial\overline{(\sigma(P)\tau(P))} = \tau(P) - \sigma(P) = \tau_0(P) - \sigma_0(P)$$

This proves (1). Next observe that

$$(\partial D_1 + D_0\partial)(\overline{PQ}) = \partial(\overline{\tau(P)\tau(Q)\sigma(Q)}) - \partial(\overline{\tau(P)\sigma(P)\sigma(Q)}) + D_0(Q) - D_0(P)$$
$$= \overline{\tau(P)\tau(Q)} + \overline{\tau(Q)\sigma(Q)} + \overline{\sigma(Q)\tau(P)} - \overline{\tau(P)\sigma(P)} - \overline{\sigma(P)\sigma(Q)}$$
$$- \overline{\sigma(Q)\tau(P)} + \overline{\sigma(Q)\tau(Q)} - \overline{\sigma(P)\tau(P)}$$

Thus

$$(\partial D_1 + D_0\partial)(\overline{PQ}) = \overline{\tau(P)\tau(Q)} - \overline{\sigma(P)\sigma(Q)}$$
$$= \tau_1(\overline{PQ}) - \sigma_1(\overline{PQ})$$

This proves (2). You should now supply a proof of (3).

The conclusion of the lemma follows immediately from these equations. For example, consider the two homomorphisms τ_1 and σ_1. In order to show that these induce the same homomorphism of $H_1(\mathscr{K})$, let c be a 1-cycle on $\mathscr{K}$. We must prove that $\tau_1(c) \sim \sigma_1(c)$. Applying (2) we find that

$$\tau_1(c) - \sigma_1(c) = \partial D_1(c) - D_0\partial(c) = \partial D_1(c)$$

since $\partial(c) = \varnothing$; thus $\tau_1(c) \sim \sigma_1(c)$.

Lemma 1 follows directly from Lemma 2. It is necessary only to verify that two simplicial approximations σ, τ of the same continuous transformation are always contiguous. Thus let P be a vertex of $\mathscr{K}$. Then by the definition of simplicial approximation,

$$\mathrm{st}(\tau(P)) \cap \mathrm{st}(\sigma(P)) \supseteq f(\mathrm{st}(P)) \neq \varnothing$$

showing that $\tau(P)$ and $\sigma(P)$ must lie on an edge together of $\mathscr{L}$. Let $\overline{PQ}$ be an edge of $\mathscr{K}$, and let R be any point of this edge except the endpoints; then

$$\mathrm{st}(\tau(P)) \cap \mathrm{st}(\sigma(P)) \cap \mathrm{st}(\tau(Q)) \cap \mathrm{st}(\sigma(Q)) \supseteq \{f(R)\} \neq \varnothing$$

showing that $\tau(P)$, $\sigma(P)$, $\tau(Q)$, $\sigma(Q)$ all lie together on a triangle of $\mathscr{L}$. The next lemma states a fundamental property of homology.

Lemma 3

Let $\mathscr{S}$, $\mathscr{T}$, and $\mathscr{U}$ be topological spaces with triangulations $\mathscr{K}$, $\mathscr{L}$, and $\mathscr{M}$, respectively. Let $f:\mathscr{S} \to \mathscr{T}$ and $g:\mathscr{T} \to \mathscr{U}$ be continuous transformations with simplicial approximations $\tau: \mathscr{K} \to \mathscr{L}$ and $\sigma:\mathscr{L} \to \mathscr{M}$, respectively. Then $\mu = \sigma\tau$ is a simplicial approximation to the continuous transformation $g \circ f:\mathscr{S} \to \mathscr{U}$, and $\mu_k = \sigma_k\tau_k$.

The proof of Lemma 3 is left as an exercise. It follows immediately that $H_k(\mu) = H_k(\sigma)H_k(\tau)$.

Exercises

1. Prove equation (3), and use it to complete the proof of Lemma 2. Incidentally find the point where the proof of Lemma 2 depends on part (b) of the definition of contiguous transformations.

2. Prove Lemma 3 by verifying that

(a) $\sigma\tau$ is a simplicial transformation

(b) $\sigma\tau$ is a simplicial approximation for $g \circ f$

(c) $\mu_k = \sigma_k \tau_k$

From Lemma 1 it follows that the induced homomorphisms $H_k(f)$ are independent of the choice of simplicial approximation when the triangulations $\mathcal{K}$ and $\mathcal{L}$ are fixed. We must now prove that $H_k(f)$ is independent of the choice of triangulations. The key is to prove that $H_k(f)$ is invariant under a subdivision of a triangulation. Figure 34.1 shows a triangulation subdivided by adding a single vertex P and two new edges a and b. This is called an **elementary subdivision**. We will consider only subdivisions that can be obtained by means of a sequence of elementary subdivisions. These include barycentric subdivisions.

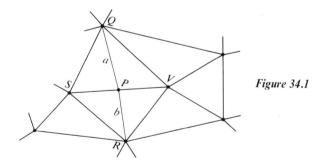

Figure 34.1

If $\mathcal{K}^+$ is a subdivision of $\mathcal{K}$, then according to the invariance theorem (§25), $\mathcal{K}$ and $\mathcal{K}^+$ have the same homology groups. Even more is true. In fact, every simplex of $\mathcal{K}$ corresponds to a chain of simplexes from $\mathcal{K}^+$ in such a way that cycles on $\mathcal{K}^+$ use the simplexes of $\mathcal{K}^+$ only in the combinations occurring in these particular chains. The clearest example of this correspondence arises in an elementary subdivision, such as in Figure 34.1. Here, for example, the 1-simplex $\overline{SV}$ of $\mathcal{K}$ corresponds to the 1-chain $\overline{SP} + \overline{PV}$ on $\mathcal{K}^+$, and the 2-simplex $\overline{SVQ}$ on $\mathscr{K}$ corresponds to the 2-chain $\overline{SPQ} + \overline{PVQ}$ on $\mathscr{K}^+$. It is this correspondence that is at the heart of the proof of the invariance principle of §25. The resulting isomorphism between $H_k(\mathscr{K})$ and $H_k(\mathscr{K}^+)$ is called the **natural isomorphism**. It arises only when one triangulation is a subdivision of the other.

The next lemma establishes invariance under subdivision, but only for the identity transformation.

Lemma 4

Let $\mathscr{K}^+$ be a subdivision of $\mathscr{K}$. Let ι be a simplicial approximation to the identity transformation from $\mathscr{K}^+$ to $\mathscr{K}$. Then $H_k(\iota)$ inverts the natural isomorphism between $H_k(\mathscr{K})$ and $H_k(\mathscr{K}^+)$. In this sense $H_k(\iota)$ is the identity homomorphism.

Consider first the case where $\mathscr{K}^+$ is an elementary subdivision of $\mathscr{K}$, as in Figure 34.1. By Lemma 1 it suffices to consider just one simplicial approximation; therefore let ι be defined by setting $\iota(U) = U$ for all vertexes of $\mathscr{K}^+$ except the added vertex P, for which we set $\iota(P) = V$. Clearly $H_k(\iota)$ is the identity on all chains except for those involving the added vertex P. To prove the lemma we must examine $H_k(\iota)$ for the particular combinations of simplexes on $\mathscr{K}^+$ that correspond to simplexes of $\mathscr{K}$. We find, for example, that $\iota_1(\overline{SP} + \overline{PV}) = \overline{SV} + \varnothing = \overline{SV}$, and $\iota_2(\overline{SPQ} + \overline{PVQ}) = \overline{SVQ}$. A systematic examination leads to the conclusion that $H_k(\iota)$ does invert the natural isomorphism between $H_k(\mathscr{K})$ and $H_k(\mathscr{K}^+)$. $H_k(\iota)$ is thus the same as the identity homomorphism of the group $H_k(\mathscr{K}) \cong H_k(\mathscr{K}^+)$.

Consider now the situation when $\mathscr{K}^+$ is any subdivision of $\mathscr{K}$. Let $\mathscr{K}_1$, $\mathscr{K}_2, \ldots, \mathscr{K}_n$ be a sequence of triangulations such that $\mathscr{K}_1 = \mathscr{K}$, $\mathscr{K}_n = \mathscr{K}^+$, and each $\mathscr{K}_i$ is an elementary subdivision of $\mathscr{K}_{i-1}$. Let $\iota_i : \mathscr{K}_i \to \mathscr{K}_{i-1}$ be a simplicial approximation to the identity transformation. By the previous paragraph, $H_k(\iota_i)$ is the identity homomorphism. By Lemma 3, $\iota = \iota_n \iota_{n-1} \cdots \iota_2$ is a simplicial approximation to the identity from $\mathscr{K}_n$ to $\mathscr{K}_1$. Since

$$H_k(\iota) = H_k(\iota_n)H_k(\iota_{n-1}) \ldots H_k(\iota_2)$$

it follows that $H_k(\iota)$ is the identity homomorphism. This proves the lemma.

PROOF OF THE INVARIANCE THEOREM

Let $\tau : \mathscr{K} \to \mathscr{L}$ and $\sigma : \mathscr{K}^* \to \mathscr{L}^*$ both be simplicial approximations of the continuous transformation $f : \mathscr{S} \to \mathscr{T}$. We must prove that $H_k(\tau) = H_k(\sigma)$.

By the Hauptvermutung there exists a triangulation $\mathcal{M}$ that is a common subdivision of $\mathcal{K}$ and $\mathcal{K}^*$ on $\mathcal{S}$. Similarly, there exists a triangulation $\mathcal{N}$ on $\mathcal{T}$ that is a common subdivision of $\mathcal{L}$ and $\mathcal{L}^*$. Now there may *not* be a simplicial approximation of f from $\mathcal{M}$ to $\mathcal{N}$, but, according to Theorem 2 of §32, there is a subdivision $\mathcal{M}^+$ of $\mathcal{M}$ on which there *is* defined a simplicial approximation of f, $\mu:\mathcal{M}^+ \to \mathcal{N}$. Naturally, $\mathcal{M}^+$, like $\mathcal{M}$, is a common subdivision of $\mathcal{K}$ and $\mathcal{K}^*$.

Let ι be a simplicial approximation of the identity from $\mathcal{M}^+$ to $\mathcal{K}$, and let ι' be a simplicial approximation of the identity from $\mathcal{N}$ to $\mathcal{L}$. Then according to Lemma 3, the simplicial transformations $\tau\iota$ and $\iota'\mu$ are both simplicial approximations for f from $\mathcal{M}^+$ to $\mathcal{L}$. Therefore by Lemma 1, $H_k(\tau\iota) = H_k(\iota'\mu)$. Now by Lemma 3, $H_k(\tau\iota) = H_k(\tau)H_k(\iota)$ and $H_k(\iota'\mu) = H_k(\iota')H_k(\mu)$; while by Lemma 4, $H_k(\iota)$ and $H_k(\iota')$ are the identity homomorphism. Putting all this together we have

$$H_k(\tau) = H_k(\tau)H_k(\iota) = H_k(\tau\iota) = H_k(\iota'\mu) = H_k(\iota')H_k(\mu) = H_k(\mu)$$

Similarly, $H_k(\sigma) = H_k(\mu)$; thus $H_k(\tau) = H_k(\sigma)$.

This proof establishes that to each continuous transformation $f:\mathcal{S} \to \mathcal{T}$ there correspond unique group homomorphisms $H_k(f):H_k(\mathcal{S}) \to H_k(\mathcal{T})$. Furthermore, from Lemmas 3 and 4 we obtain the following corollaries.

Corollary 1

If $f:\mathcal{S} \to \mathcal{S}$ is the identity function, then $H_k(f)$ is the identity homomorphism.

Corollary 2

If $f:\mathcal{S} \to \mathcal{T}$ and $g:\mathcal{T} \to \mathcal{U}$ are continuous transformations, then $H_k(g \circ f) = H_k(g)H_k(f)$.

Exercises

3. Show how barycentric subdivision can be obtained from a sequence of elementary subdivisions.

4. Complete the proof of Lemma 4.

---

§35 MATRIXES

The invariance theorem establishes that properties of the group homomorphisms $H_k(f)$ will reflect properties of the continuous transformation f. In order to draw out the information $H_k(f)$ can supply concerning f, we are forced to undertake a study of group homomorphisms. This is the purpose of this section.

Let h be a homomorphism from a group G to a group K. Let n be the rank of G and let $\{x_1, \ldots, x_n\}$ be a basis for G. This basis will be called a **proper basis** if every element x in G can be uniquely expressed

$$x = a_1 x_1 + a_2 x_2 + \cdots + a_n x_n + t$$

where the coefficients $a_1, a_2, \ldots, a_n$ are integers and t is a torsion element. The existence of such a basis is guaranteed by the direct sum theorem. The action of h on G is determined by its action on the basis elements $\{x_1, \ldots, x_n\}$ and by its action on the torsion subgroup of G. For this reason proper bases are the only bases of interest in the study of homomorphisms, so we can occasionally drop the word proper without fear of confusion.

Let m be the rank of K and let $\{y_1, y_2, \ldots, y_m\}$ be a proper basis for H. Then every element y in K can be written uniquely as

$$y = b_1 y_1 + b_2 y_2 + \ldots + b_m y_m + s$$

where the coefficients $b_1, \ldots, b_m$ are integers and s is a torsion element in K. Applying this to the elements $\{h(x_1), h(x_2), \ldots, h(x_n)\}$ we obtain the array

$$h(x_1) = b_{1,1} y_1 + b_{2,1} y_2 + \cdots + b_{m,1} y_m + s_1$$
$$h(x_2) = b_{1,2} y_1 + b_{2,2} y_2 + \cdots + b_{m,2} y_m + s_2$$
$$\vdots$$
$$h(x_n) = b_{1,n} y_1 + b_{2,n} y_2 + \cdots + b_{m,n} y_m + s_n$$

Ignoring the torsion elements, which play no role in the applications we contemplate, the homomorphism h is determined by the coefficients $b_{i,j}$. Thus we shall expect to obtain information about the homomorphism h from the rectangular matrix

$$B = \begin{bmatrix} b_{1,1} & b_{1,2} & \cdots & b_{1,n} \\ b_{2,1} & b_{2,2} & \cdots & b_{2,n} \\ \vdots & & & \\ b_{m,1} & b_{m,2} & \cdots & b_{m,n} \end{bmatrix}$$

We abbreviate B by writing $B = [[b_{i,j}]]$. The matrix B is called the **matrix of h** with respect to the bases $\{x_1, x_2, \ldots, x_n\}$ and $\{y_1, y_2, \ldots, y_m\}$.

Examples

The simplest nontrivial examples involve the group of two-dimensional integers $\mathscr{L}^2$. Let $\{a, b\}$ be any proper basis for $\mathscr{L}^2$. Let h be a homomorphism from $\mathscr{L}^2$ to $\mathscr{L}^2$. Using $\{a, b\}$ as basis for both the range and domain of h, let

$$h(a) = b_{1,1}a + b_{2,1}b$$
$$h(b) = b_{1,2}a + b_{2,2}b$$

Then the matrix of h with respect to this choice of bases is

$$B = \begin{bmatrix} b_{1,1} & b_{1,2} \\ b_{2,1} & b_{2,2} \end{bmatrix}$$

Conversely, every such matrix B determines a homomorphism of the group $\mathscr{L}^2$ by means of the formula

$$h(x) = (b_{1,1}m + b_{1,2}n)a + (b_{2,1}m + b_{2,2}n)b$$

where $x = ma + nb$. In this case the matrix completely determines the homomorphism, because there is no torsion.

Let $\mathscr{S}$ be a torus. The first homology group of $\mathscr{S}$, $H_1(\mathscr{S})$, is the group $\mathscr{L}^2$. Therefore if f is a continuous transformation from $\mathscr{S}$ to $\mathscr{S}$, the induced homomorphism $H_1(f)$ is a group homomorphism from $\mathscr{L}^2$ to $\mathscr{L}^2$. Let the surface $\mathscr{S}$ be given by its plane model (Figure 35.1a); then the edges a and b form a basis for $H_1(\mathscr{S})$. Let $B = [[b_{i,j}]]$ $(i, j = 1, 2)$ be any (2×2) matrix. Let us describe how to construct a continuous transformation $f : \mathscr{S} \to \mathscr{S}$ such that the matrix of $H_1(f)$ with respect to the basis $\{a, b\}$ of $H_1(\mathscr{S})$ is B. We start by imbedding $\mathscr{S}$ in its universal covering surface $\mathscr{P}$, which is a plane (Figure 35.1b). The points (x, y) of $\mathscr{P}$ are then transformed by the linear transformation: $B(x, y) = (b_{1,1}x + b_{1,2}y, b_{2,1}x + b_{2,2}y)$. Figure 35.1c illustrates this with the matrix

$$B = \begin{bmatrix} 2 & 1 \\ 1 & 1 \end{bmatrix}$$

Finally we apply the covering transformation to bring the image of our transformation back inside $\mathscr{S}$ (Figure 35.1d). Let the transformation f_B be the composition of these three operations: imbedding $\mathscr{S}$ in the covering surface, the transformation B of $\mathscr{P}$, and the covering transformation $\mathscr{P} \to \mathscr{S}$.

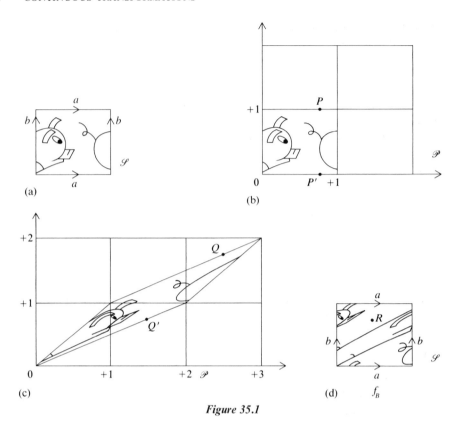

Figure 35.1

In order to prove that f_B is well defined, we must prove that identified points on the edges a and b always end up covering the same point of $\mathscr{S}$ after being transformed by B. For example, the points P, P' in Figure 35.1b are transformed to the points Q, Q' in Figure 35.1c, both of which cover the point R of Figure 35.1d. You can easily supply this proof by means of similar triangles.

In order to compute the induced homomorphisms $H_k(f_B)$ we need a simplicial approximation for f_B. This is easy to obtain in this case because f_B is essentially the linear transformation B, which maps straight lines to straight lines and triangles to triangles. We start with a triangulation of the image of $\mathscr{S}$ in $\mathscr{P}$. This is obtained by dividing $\mathscr{P}$ into parallelograms with sides parallel to the sides of the image of $\mathscr{S}$ passing through the lattice points $\mathscr{L}^2$ in $\mathscr{P}$. These parallelograms are then divided into triangles by drawing diagonals. Figure 35.2b gives an example for the matrix

$$B = \begin{bmatrix} 1 & 2 \\ -1 & 1 \end{bmatrix}$$

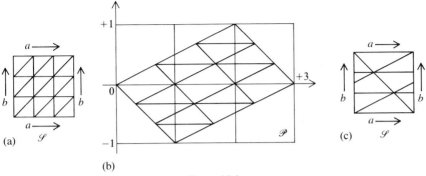

Figure 35.2

This triangulation of $\mathscr{P}$ gives rise to triangulations in $\mathscr{S}$ (Figure 35.2a) *and in* $f_B(\mathscr{S})$ (Figure 35.2c). With respect to these triangulations, f_B is its own simplicial approximation. Let h_B be the homomorphism $H_1(f_B)$. In the example of Figure 35.2 it is clear that $h_B(a)$ is homologous to $a - b$ (Figure 35.3a), while $h_B(b)$ is homologous to $2a + b$ (Figure 35.3b). In the same way it is clear in general that

$$h_B(a) = b_{1,1}a + b_{2,1}b$$
$$h_B(b) = b_{1,2}a + b_{2,2}b$$

Thus the matrix of $H_1(f_B)$ is B.

The transformations f_B are useful examples of nontrivial continuous transformations. Notice that they need not be one-to-one. In fact the transformation in Figure 35.2 covers the torus exactly three times. Thus you will note that the triangulation of $\mathscr{S}$ in Figure 35.2a contains 18 triangles, each of which covers exactly one of the 6 triangles in Figure 35.2c.

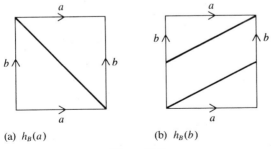

(a) $h_B(a)$ (b) $h_B(b)$

Figure 35.3

Exercises

1. Let G be a group and let $\{x_1, x_2, \ldots, x_n\}$ and $\{y_1, y_2, \ldots, y_n\}$ be *two* proper bases for G. Then each basis can be expressed uniquely in terms of the other:

$$x_1 = a_{1,1}y_1 + a_{2,1}y_2 + \cdots + a_{n,1}y_n$$
$$\vdots$$
$$x_n = a_{1,n}y_1 + a_{2,n}y_2 + \cdots + a_{n,n}y_n$$

and

$$y_1 = b_{1,1}x_1 + b_{2,1}x_2 + \cdots + b_{n,1}x_n$$
$$\vdots$$
$$y_n = b_{1,n}x_1 + b_{2,n}x_2 + \cdots + b_{n,n}x_n$$

Show that the matrixes $A = [[a_{i,j}]]$ and $B = [[b_{i,j}]]$ are inverses, that is, $AB = BA = I$, where AB is the usual product of matrixes $AB = [[p_{i,j}]]$,

$$p_{i,k} = a_{i,1}b_{1,k} + a_{i,2}b_{2,k} + \cdots + a_{i,n}b_{n,k} = \sum_{j=1}^{n} a_{i,j}b_{j,k}$$

and I is the identity matrix,

$$I = \begin{bmatrix} 1 & 0 & 0 & \cdots & 0 \\ 0 & 1 & 0 & \cdots & 0 \\ \vdots & & & & \\ 0 & 0 & 0 & \cdots & 1 \end{bmatrix}$$

2. Let h be a group homomorphism from a group G to itself (technically h is called an **automorphism**). Let $\{x_1, x_2, \ldots, x_n\}$ and $\{y_1, y_2, \ldots, y_n\}$ be two bases for G related by the matrixes A and B of the preceeding exercise. Let $H = [[h_{i,j}]]$ be the matrix of h with respect to the basis $\{x_1, x_2, \ldots, x_n\}$. Show that the matrix of h with respect to $\{y_1, y_2, \ldots, y_n\}$ is AHB.

3. Let $B = [[b_{i,j}]]$ be a (2×2) matrix, and left f_B be the corresponding continuous transformation of the torus as defined above. Show by examples that f_B can cover the torus any whole number of times. Find examples to show that f_B can reverse the orientation of the torus.

4. Show that the area of the parallelogram that arises in the definition of the transformation f_B (Figure 35.1c and Figure 35.2b) is equal to $|\det(B)|$, where $\det(B)$, the **determinant** of B, equals $b_{1,1}b_{2,2} - b_{1,2}b_{2,1}$. Show that the sign of $\det(B)$ is plus or minus according as f_B preserves or reverses the

orientation of the torus. Conclude that the matrix for the linear transformation $H_2(f_B)$ is the number $\det(B)$.

5. Let r_A be a rotation of the torus. Find matrixes for the homomorphisms $H_k(r_A)$.

6. Find matrixes for homology homomorphisms induced by the following transformations of the complex sphere:

(a) $f(z) = z^n$ (b) $f(z) = z^{-n}$

(c) a Möbius transformation   (d) $f(z) = \bar{z}^n$

7. Show that all transformations of the projective plane have the same matrixes for their homology homomorphisms.

8. Use the universal covering surface of the Klein bottle to construct transformations of the Klein bottle, and compute the matrixes of the corresponding homology homomorphisms.

Let $f : \mathscr{S} \to \mathscr{T}$ be a continuous transformation. When bases are chosen for the homology groups of $\mathscr{S}$ and $\mathscr{T}$, the three homomorphisms $H_k(f)$ determine three matrixes from which we might expect to be able to determine properties of the transformation f. However, there is an invariance problem here (there seems to be one everywhere in topology). These matrixes depend on the *choice* of the bases for the homology groups. A change in bases produces a definite change in the matrixes. Exercise 2 shows how this change comes about in the special case of a group automorphism. The matrixes $H_k(f)$ are thus *not* independent of the choice of basis: there is no invariance theorem for these matrixes. We must discover features of the matrixes that are independent of the choice of basis. These features can then be used to determine properties of the continuous transformation f.

In one simple case this invariance problem almost disappears. Let h be a homomorphism between groups G and K that are isomorphic to the rank-one group $\mathscr{Z}$. Since $\mathscr{Z}$ has only two proper bases ± 1, it follows that the matrix of h is a number *whose sign alone is affected by a change in basis*. In this case the absolute value of the matrix element of h is independent of the choice of basis.

Let us look for examples of the group $\mathscr{Z}$ among the homology groups in order to exploit this limited invariance. Springing first to mind are the zero-dimensional groups $H_0(\mathscr{S})$, which, for all connected spaces $\mathscr{S}$, are isomorphic to $\mathscr{Z}$. The matrix $H_0(f)$, however, is quite uninteresting (Exercise 9). Another example that lies immediately before us is the group $H_2(\mathscr{S})$, which for compact, connected, *orientable* surfaces is isomorphic to $\mathscr{Z}$. This example leads to the following definition.

Definition

Let $f:\mathscr{S} \to \mathscr{T}$ be a continuous transformation between two compact, connected, orientable surfaces. The number b that constitutes the (1×1) matrix for the homomorphism $H_2(f)$ is called the **Brouwer degree** *of f.*

Intuitively the Brouwer degree counts the number of times the transformation f covers $\mathscr{T}$ with $\mathscr{S}$. The Brouwer degree is determined only up to a change of sign. However, if the two spaces are the same, then there is only one basis to choose, and it turns out that no matter whether $+1$ or -1 is chosen, the matrix of $H_2(f)$ is the same. In this case the sign of the Brouwer degree has a meaning, too: indicating whether or not f reverses the orientation of $\mathscr{S}$. For example, consider the polynomial transformation $p(z) = z^2$ of the complex sphere. Under this transformation the sphere is covered twice by itself. The Brouwer degree on this account is 2. On the other hand, the Brouwer degree of the transformation $f(z) = \bar{z}^2$ is -2, because here z^2 is combined with the reflection $\bar{z}$, which reverses the orientation of the sphere.

WINDING NUMBERS

The idea of the Brouwer degree can be applied one dimension lower to give a new definition of the winding number. Let V be a vector field on a Jordan curve $\mathscr{J}$. $\mathscr{J}$ is topologically equivalent to a circle and therefore has the same homology groups as a circle. Let $\mathscr{J}^*$ be a circle, and let us associate to each point P of $\mathscr{J}$ the point $f(P)$ on $\mathscr{J}^*$ at which the vector $V(P)$ points if placed at the center of $\mathscr{J}^*$. This defines a continuous transformation f from $\mathscr{J}$ to $\mathscr{J}^*$. This transformation in turn induces homomorphisms from the homology groups of $\mathscr{J}$ to those of $\mathscr{J}^*$. In this case, $H_1(\mathscr{J}) = H_1(\mathscr{J}^*) = \mathscr{Z}$, so that the matrix of $H_1(f)$ consists of a single number W. This number W, the one-dimensional **Brouwer degree of f**, is actually the winding number of V on $\mathscr{J}$. To prove this, let $\mathscr{J}^*$ be divided into three edges by the points A, B, and C shown in Figure 35.4. At the same time let $\mathscr{J}$ also be divided into a complex $\mathscr{K}$. Recall that the winding number as defined in §8 is found by labeling the vertexes P on $\mathscr{J}$ according to the direction of the vector $V(P)$. For each point P of $\mathscr{K}$, let $l(P)$ be the appropriate label A, B, or C, according to the conventions laid down in §8. The winding number is the algebraic sum of the numbers of edges labeled AB counted by orientation. The combinatorial connection between the vector field V and the transformation f is that if the complex $\mathscr{K}$ is chosen fine enough, then *the labels $l(P)$ form a simplicial approximation to f.* This can be proven by a compactness argument. The Brouwer degree of f counts the number of times f covers the circle $\mathscr{J}^*$. In

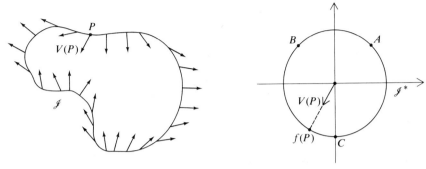

Figure 35.4

particular, W counts the number of times f (or its simplicial approximation l) covers the segment AB of $\mathcal{I}^*$. Thus W equals the winding number of V on $\mathcal{I}$.

Note the close analogy between the concepts of Brouwer degree and winding number. They appear as two- and one-dimensional versions of the same homological idea. Each is a single number that represents the number of times a transformation f covers its range space. Each can be obtained from the matrix of an induced homology group homomorphism.

Exercises

9. Let $f:\mathcal{S} \to \mathcal{T}$ be a continuous transformation between connected spaces. Show that the matrix of $H_0(f)$ is always the single number ± 1. In case $\mathcal{S} = \mathcal{T}$, show that the matrix of $H_0(f)$ is always $+1$ when the same basis is used for domain and range.

10. Supply the compactness argument needed to complete the proof that the winding number equals the one-dimensional Brouwer degree.

11. Use the fact that the winding number equals a Brouwer degree to prove that the winding number does not depend on

(a) the choice of the particular direction due north

(b) the use of the first and second quadrants as the labeling regions A and B instead of any other adjacent sectors of the plane about the origin

(c) the number of labels used (which can be any number).

This result fills a gap left back in §8.

12. Let $f:\mathcal{S} \to \mathcal{T}$ and $g:\mathcal{T} \to \mathcal{U}$ be continuous transformations between compact, connected, orientable surfaces. Show that the Brouwer degree of $g \circ f$ equals the product of the Brouwer degrees of f and g.

13. Find the Brouwer degrees for the following transformations:

(a) the rotation r_A of the torus

(b) the transformation f_B of the torus

(c) the polynomial $p(z) = z^n$ on the complex sphere

(d) $f(z) = z^{-n}$ on the complex sphere

---

§36 THE LEFSCHETZ FIXED POINT THEOREM

In case the homomology groups are not of rank one, we must find some feature of the matrixes invariant under a change in basis in order to extract something of geometric significance from the homology homomorphisms. One such feature is the trace of a square matrix. This applies only when the domain and range of the homomorphism are the same (i.e., the homomorphism is an automorphism) and the same basis is used for range and domain. Thus the topological applications of the trace are limited to continuous transformations of a space into itself. We will use the trace to investigate fixed points of such transformations, bringing us back to one of the first topics discussed in this book. We shall prove the Lefschetz fixed point theorem, one of the most important theorems in topology: a simultaneous generalization of the Brouwer fixed point theorem, the Euler characteristic, and the Poincaré index theorem.

Definition

*Let $B = [[b_{i,j}]]$ be a square matrix, $1 \le i, j \le n$. The **trace of B** is defined by*

$$\mathrm{tr}(B) = b_{1,1} + b_{2,2} + \cdots + b_{n,n}$$

In words, $\mathrm{tr}(B)$ is the sum of the diagonal entries of B. The diagonal entries measure the tendency of the homomorphism to leave the elements of the basis fixed. Thus it is not surprising that the trace appears in applications to fixed point theorems. Note that when the matrix B is (1×1), then the trace equals the single entry in B, and if B is the identity matrix, then $\mathrm{tr}(B) = n$.

Let $A = [[a_{i,j}]]$ be a second $(n \times n)$ matrix. The product

$$AB = \left[\left[\sum_k a_{i,k} b_{k,j}\right]\right]$$

has trace

$$\mathrm{tr}(AB) = \sum_{ik} a_{i,k} b_{k,i}$$

Thus $\operatorname{tr}(AB) = \operatorname{tr}(BA)$. This is the principle algebraic property of the trace.

As an application, suppose that $H = [[h_{i,j}]]$ is the matrix of a group homomorphism $h: G \to G$. Let two bases be chosen for G. These bases are related by two matrixes A and B that are inverses: $BA = I$ (Exercise 1 of §35). The matrix of h with respect to the second basis is AHB (Exercise 2 of §35). Observe now that

$$\operatorname{tr}(AHB) = \operatorname{tr}(A(HB)) = \operatorname{tr}((HB)A) = \operatorname{tr}(HBA) = \operatorname{tr}(H)$$

Thus the trace of the matrix H is the same for *all* matrixes representing the homomorphism h, independent of the choice of basis! This common value is called the trace of the homomorphism itself, written $\operatorname{tr}(h)$. Note that if h is the identity homomorphism, then $\operatorname{tr}(h) = \operatorname{rank}(G)$.

Given a continuous transformation f of a topological space $\mathscr{S}$ into itself, we may apply the trace to homomorphisms both of homology and chain groups. Thus let $\mathscr{K}$ be a triangulation of $\mathscr{S}$ and let $\mathscr{K}^+$ be a subdivision of $\mathscr{K}$ on which there exists a simplicial approximation τ of f. The simplicial map τ induces homomorphisms τ_k from $C_k(\mathscr{K}^+)$ to $C_k(\mathscr{K})$. Every chain on $\mathscr{K}$, however, can be considered a chain on $\mathscr{K}^+$ by writing out every simplex on $\mathscr{K}$ as the sum of the simplexes into which it is subdivided. Therefore we may consider τ_k as a homomorphism of $C_k(\mathscr{K}^+)$ to itself, in which case the trace of τ_k is defined. We have also the traces of the automorphisms $h_k = H_k(f)$ of the homology groups $H_k(\mathscr{S})$. These six traces are linked together by the **Hopf trace formula**:

$$\operatorname{tr}(\tau_0) - \operatorname{tr}(\tau_1) + \operatorname{tr}(\tau_2) = \operatorname{tr}(h_0) - \operatorname{tr}(h_1) + \operatorname{tr}(h_2) \tag{1}$$

To prove this important relation requires careful choice of bases in the chain groups. Beginning with the group C_2, we first choose a basis for the subgroup Z_2 of 2-cycles: $\{z_1, z_2, \ldots, z_m\}$. Then choose additional elements $\{y_1, y_2, \ldots, y_n\}$ so that $\{z_1, z_2, \ldots, z_m, y_1, y_2, \ldots, y_n\}$ is a basis for C_2. With respect to this basis we have

$$
\begin{aligned}
\tau_2(z_1) &= a_{1,1}z_1 + a_{2,1}z_2 + \cdots + a_{m,1}z_m \\
\tau_2(z_2) &= a_{1,2}z_1 + a_{2,2}z_2 + \cdots + a_{m,2}z_m \\
&\ \vdots \\
\tau_2(z_m) &= a_{1,m}z_1 + a_{2,m}z_2 + \cdots + a_{m,m}z_m \\
\tau_2(y_1) &= b_{1,1}z_1 + b_{2,1}z_2 + \cdots + b_{m,1}z_m + c_{1,1}y_1 + c_{2,1}y_2 + \cdots + c_{n,1}y_n \\
\tau_2(y_2) &= b_{1,2}z_1 + b_{2,2}z_2 + \cdots + b_{m,2}z_m + c_{1,2}y_1 + c_{2,2}y_2 + \cdots + c_{n,2}y_n \\
&\ \vdots \\
\tau_2(y_n) &= b_{1,n}z_1 + b_{2,n}z_2 + \cdots + b_{m,n}z_m + c_{1,n}y_1 + c_{2,n}y_2 + \cdots + c_{n,n}y_n
\end{aligned}
\tag{2}
$$

Thus if $A = [[a_{ij}]]$, $B = [[b_{i,j}]]$, and $C = [[c_{i,j}]]$, then the matrix of τ_2 takes the following block form:

$$\tau_2 = \left[\begin{array}{c|c} \overset{\displaystyle m}{\overbrace{A}} & \overset{\displaystyle n}{\overbrace{B}} \\ \hline 0 & C \end{array}\right] \begin{array}{l}]\ m \\ ]\ n \end{array}$$

Note the block of zeros in the lower left corner. This occurs because τ_2 transforms 2-cycles into 2-cycles; therefore the expansion of the element $\tau_2(z_i)$ doesn't involve the elements $\{y_j\}$.

Turning to 1-chains, we begin with the elements $x_1 = \partial(y_1), x_2 = \partial(y_2), \ldots, x_n = \partial(y_n)$. They form a basis for the subgroup B_1 of 1-boundaries (Exercise 2). Next choose elements $w_1, w_2, \ldots, w_q$ so that $\{x_1, \ldots, x_n, w_1, w_2, \ldots, w_q\}$ is a basis for the group Z_1 of 1-cycles. Finally choose elements $t_1, t_2, \ldots, t_p$ to complete a basis $\{x_1, \ldots, x_n, w_1, \ldots, w_q, t_1, \ldots, t_p\}$ for C_1. With respect to this basis we have

$$\tau_1(x_1) = d_{1,1}x_1 + \cdots + d_{n,1}x_n$$
$$\vdots$$
$$\tau_1(x_n) = d_{1,n}x_1 + \cdots + d_{n,n}x_n$$
$$\tau_1(w_1) = e_{1,1}x_1 + \cdots + e_{n,1}x_n + g_{1,1}w_1 + \cdots + g_{q,1}w_q$$
$$\vdots \tag{3}$$
$$\tau_1(w_q) = e_{1,q}x_1 + \cdots + e_{n,q}x_n + g_{1,q}w_1 + \cdots + g_{q,q}w_q$$
$$\tau_1(t_1) = f_{1,1}x_1 + \cdots + f_{n,1}x_n + h_{1,1}w_1 + \cdots + h_{q,1}w_q + j_{1,1}t_1 + \cdots + j_{p,1}t_p$$
$$\vdots$$
$$\tau_1(t_p) = f_{1,p}x_1 + \cdots + f_{n,p}x_n + h_{1,p}w_1 + \cdots + h_{q,p}w_q + j_{1,p}t_1 + \cdots + j_{p,p}t_p$$

Thus if $D = [[d_{i,j}]], E = [[e_{i,j}]], F = [[f_{i,j}]], G = [[g_{i,j}]], H = [[h_{i,j}]]$, and $J = [[j_{i,j}]]$, then the matrix of τ_1 takes on the block form

$$\tau_1 = \left[\begin{array}{c|c|c} \overset{\displaystyle n}{\overbrace{D}} & \overset{\displaystyle q}{\overbrace{E}} & \overset{\displaystyle p}{\overbrace{F}} \\ \hline 0 & G & H \\ \hline 0 & 0 & J \end{array}\right] \begin{array}{l}]\ n \\ ]\ q \\ ]\ p \end{array}$$

The blocks of zeros occur because B_1 and Z_1 are subgroups of C_1.

For the last step we observe that the elements $v_1 = \partial(t_1), v_2 = \partial(t_2), \ldots, v_p = \partial(t_p)$ form a basis for the group B_0 of zero-boundaries. Choose elements $u_1, u_2, \ldots, u_r$ to fill out a basis $\{v_1, \ldots, v_p, u_1, \ldots, u_r\}$ for the group C_0. With respect to this basis,

$$\tau_0(v_1) = k_{1,1}v_1 + \cdots + k_{p,1}v_p$$

$$\vdots$$

$$\tau_0(v_p) = k_{1,p}v_1 + \cdots + k_{p,p}v_p$$

$$\tau_0(u_1) = l_{1,1}v_1 + \cdots + l_{p,1}v_p + m_{1,1}u_1 + \cdots + m_{r,1}u_r \qquad (4)$$

$$\vdots$$

$$\tau_0(u_r) = l_{1,r}v_1 + \cdots + l_{p,r}v_p + m_{1,r}u_1 + \cdots + m_{r,r}u_r$$

Thus if $K = [[k_{i,j}]]$, $L = [[l_{i,j}]]$, and $M = [[m_{i,j}]]$, then the matrix of τ_0 takes on the block form

$$\tau_0 = \left[\begin{array}{c|c} \overbrace{K}^{p} & \overbrace{L}^{r} \\ \hline 0 & M \end{array} \right] \begin{array}{l}] \ p \\ ] \ r \end{array}$$

These block forms clearly imply

$$\mathrm{tr}(\tau_0) = \mathrm{tr}(K) + \mathrm{tr}(M)$$

$$\mathrm{tr}(\tau_1) = \mathrm{tr}(D) + \mathrm{tr}(G) + \mathrm{tr}(J)$$

$$\mathrm{tr}(\tau_2) = \mathrm{tr}(A) + \mathrm{tr}(C)$$

Consider next the question of matrixes for the homology transformations. Recall (from §33) that the homology transformation h_k is obtained by restricting the simplicial transformation τ_k to the subgroup of k-cycles and replacing equality by homology. Take the transformation τ_1, for example. Restricting τ_1 to 1-cycles means that we ignore the last lines of equation (3), the lines that involve the elements $t_1, \ldots, t_p$. Then replacing equality by homology, the elements $w_1, \ldots, w_q$ form a homology basis for the 1-cycles so that

$$\tau_1(w_1) \sim g_{1,1}w_1 + \cdots + g_{q,1}w_q$$

$$\vdots$$

$$\tau_1(w_q) \sim g_{1,q}w_1 + \cdots + g_{q,q}w_q$$

Therefore the matrix of h_1 with respect to the homology basis $w_1, \ldots, w_q$ is G. Similarly, we find that the matrix of h_2 is A, while the matrix of h_0 is M.

Now it is easy to show, using equations (1) and (2) from §33, that $C = D$ and $K = J$. Putting all these results together yields

$$\mathrm{tr}(\tau_0) - \mathrm{tr}(\tau_1) + \mathrm{tr}(\tau_2) = \mathrm{tr}(M) - \mathrm{tr}(G) + \mathrm{tr}(A)$$

$$= \mathrm{tr}(h_0) - \mathrm{tr}(h_1) + \mathrm{tr}(h_2)$$

concluding the proof.

Exercises

1. Prove that $\text{tr}(AB) = \text{tr}(BA)$.
2. Justify the block forms shown for τ_1 and τ_0. Explain why $C = D$ and $K = J$.
3. Show that the quantities m, q, and r used in the proof of the Hopf trace formula are the Betti numbers of $\mathscr{S}$.
4. Deduce the Euler-Poincaré formula as a special case of the Hopf trace formula.
5. Which lowercase Roman letters were *not* used in the proof of the Hopf trace formula? Which were used *twice*, that is to say, in two different senses?

INDEX OF A FIXED POINT

This definition will have a familiar sound. It is based on the idea of replacing a transformation by a vector field that was used way back in Chapter Two.

Definition

*Let f be a continuous transformation from a surface $\mathscr{S}$ to itself. Let P be a fixed point of f. We assume that all fixed points are isolated; therefore P has a neighborhood $\mathscr{N}$, topologically equivalent to a disk, in which P is the only fixed point of f. Altering the surface by a topological transformation if necessary, we can suppose that $\mathscr{N}$ is a disk. Since P is a fixed point, there will be a smaller disk $\mathscr{D}$ containing P such that $f(\mathscr{D}) \subseteq \mathscr{N}$. Then the transformation f defines a vector field V on $\mathscr{D}$ in which $V(Q)$ (for Q in $\mathscr{D}$) is the vector from Q to $f(Q)$. The **index of f at P** is defined to be the winding number of this vector field around the circumference of the disk $\mathscr{D}$.*

In other words, the index $I(P)$ is the index of P as a critical point of the vector field V. The application of this definition is illustrated in Figure 36.1. The transformation is $f(z) = z^2$, and there are two fixed points, 0 and 1, both of index one.

For the proof of the Lefschetz fixed point theorem, the index will be computed by a different method, one that links the index with homology. Unfortunately this alternative method works only for a special type of fixed point. The fixed point is called **normal** when the disk $\mathscr{D}$ of the preceding definition can be chosen in such a way that the vectors $V(Q)$ on the boundary

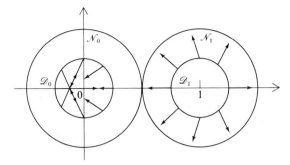

Figure 36.1

of $\mathscr{D}$ all point outside $\mathscr{D}$. In other words, the transformation f stretches the boundary of $\mathscr{D}$ outside $\mathscr{D}$ itself. In Figure 36.1, 1 is a normal fixed point but 0 is not. For normal fixed points the index can be computed as follows. The surface $\mathscr{S}$ is enlarged by adding on a disk $\mathscr{D}'$ whose points are in one-to-one correspondence with those of the given disk $\mathscr{D}$; that is, for every point Q of $\mathscr{D}$ there is a matching point Q' on $\mathscr{D}'$. The cell $\mathscr{D}'$ is attached to $\mathscr{S}$ by identifying $\mathscr{D}$ and $\mathscr{D}'$ at their corresponding boundary points (Figure 36.2). Thus $\mathscr{D}$ and $\mathscr{D}'$ together form a sphere, $\mathscr{C}$, attached to the surface $\mathscr{S}$ at its equator. Let $\mathscr{S}'$ denote the resulting topological space. $\mathscr{S}'$ *is not a surface.* The transformation f can be extended to a continuous transformation f' of $\mathscr{S}'$ by defining

$$f'(Q') = \begin{cases} f(Q)' & \text{if } f(Q) \text{ is in } \mathscr{D} \\ f(Q) & \text{if } f(Q) \text{ isn't in } \mathscr{D} \end{cases} \tag{5}$$

Thus f' maps the upper hemisphere $\mathscr{D}'$ in a manner that is a reflection of the mapping f on the lower hemisphere $\mathscr{D}$. The extended transformation f' is continuous, but *only* because P is a normal fixed point.

In computing the second homology group, it turns out that no matter how $\mathscr{S}'$ is triangulated, the sphere $\mathscr{C}$ forms a new 2-cycle not present on $\mathscr{S}$. Let

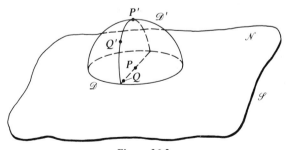

Figure 36.2

$\{\mathscr{C}, C_1, C_2, \ldots, C_n\}$ be a basis for $H_2(\mathscr{S}')$ containing $\mathscr{C}$. Under the homomorphism $h_2' = H_2(f')$ we have

$$h_2'(\mathscr{C}) = \beta\mathscr{C} + \beta_1 C_1 + \beta_2 C_2 + \cdots + \beta_n C_n$$

The coefficient β is the index of f at P, $I(P)$. The proof of this result is similar to the treatment of winding numbers in §35. The two methods of computing the index are linked by the fact that the labeling of points on $\mathscr{D}$ according to the direction of the vectors $V(Q)$ can be used as a simplicial approximation to the function f' on $\mathscr{C}$ along the equator. The intuitive significance of this new method for computing the index is that $I(P)$ is now seen as measuring the number of times f covers the neighborhood $\mathscr{D}$ of P with itself. Since the only difference between $H_2(\mathscr{S})$ and $H_2(\mathscr{S}')$ is the cycle $\mathscr{C}$, the index can also be written

$$I(P) = \text{tr}(h_2') - \text{tr}(h_2) \tag{6}$$

If $P_1, P_2, \ldots, P_n$ are all isolated normal fixed points of f, let $\mathscr{D}_1, \mathscr{D}_2, \ldots,$ $\mathscr{D}_n$ be suitable neighborhoods and let $\mathscr{S}'$ be the surface obtained by attaching copies of these neighborhoods to $\mathscr{S}$ forming n spheres $\mathscr{C}_1, \mathscr{C}_2, \ldots, \mathscr{C}_n$. Extending f to a transformation f' of $\mathscr{S}'$ as before, we have in place of equation (6),

$$\text{tr}(h_2') - \text{tr}(h_2) = I(P_1) + I(P_2) + \cdots + I(P_n) \tag{7}$$

Exercises

6. Find the indexes of the fixed points of the following transformations of the plane:

(a) $f(z) = z^3$ (b) $f(z) = z^4$

(c) $f(z) = z^n$ (d) $f(z) = z^2 + z$

(e) $f(z) = z^3 + z$ (f) $f(z) = z/(z+1)$

7. Let f be a transformation of the complex sphere with a fixed point at ∞. Let $t(z) = 1/z$. Explain why the index of f at ∞ is equal to the index of $g = t \circ f \circ t$ at zero. Use this device to find the index at ∞ for the following transformations:

(a) $f(z) = z^2$ (b) $f(z) = z^n$

(c) $f(z) = z^2 + z$ (d) $f(z) = z + 1$

8. Prove that the two methods of computing the index of a fixed point are equivalent.

We now come to the main theorem of this chapter. Let f be a continuous transformation of a compact surface $\mathscr{S}$ into itself. Let $h_k = H_k(f)$ be the induced homology homomorphism on $H_k(\mathscr{S})$, $k = 0, 1, 2$. The number

$$L(f) = \text{tr}(h_0) - \text{tr}(h_1) + \text{tr}(h_2)$$

is called the **Lefschetz number** of f. It acts as a sort of Euler characteristic for the transformation f.

The Lefschetz Fixed Point Theorem

Suppose that f has only a finite number of fixed points, $P_1, P_2, \ldots, P_n$. Then

$$I(P_1) + I(P_2) + \cdots + I(P_n) = L(f) \tag{9}$$

Equation (9) is called the **Lefschetz fixed point formula**.

The following proof is due to Hopf [14]. Suppose first that f has *no* fixed points. We must prove that $L(f) = 0$. Suppose on the contrary that $L \neq 0$. Let $\mathscr{K}$ be any triangulation of $\mathscr{S}$, and let τ be a simplicial approximation to f on $\mathscr{K}$. According to the Hopf trace formula, $\text{tr}(\tau_0) - \text{tr}(\tau_1) + \text{tr}(\tau_2)$ is not equal to zero. Therefore at least one triangle A of $\mathscr{K}$ has the property that $f(A)$ intersects A. By taking a sequence of triangulations $\mathscr{K}_n$ of smaller and smaller mesh, for example a sequence of barycentric subdivisions, we obtain a sequence of triangles A_n such that $f(A_n) \cap A_n \neq \varnothing$. By a compactness argument, it follows that f *has* a fixed point. This contradiction proves the theorem in this case.

Turning to the general case, let $P_1, P_2, \ldots, P_n$ be the fixed points of f. We assume that all fixed points are normal. Any nonnormal fixed point P can be made normal by composing f with a stretching transformation around P. This stretching will not change the number of fixed points of f or their indexes. Therefore the conclusion of the theorem once proven for the altered transformation will apply as well to the original transformation. Let $\mathscr{D}_1$, $\mathscr{D}_2, \ldots, \mathscr{D}_n$ be suitable neighborhoods of the fixed points, and let $\mathscr{D}_1'$, $\mathscr{D}_2', \ldots, \mathscr{D}_n'$ be copies of these neighborhoods attached along their boundaries to the disks $\mathscr{D}_1, \mathscr{D}_2, \ldots, \mathscr{D}_n$ forming a new space $\mathscr{S}'$ containing the n spheres $\mathscr{C}_1 = \mathscr{D}_1 \cup \mathscr{D}_1', \ldots, \mathscr{C}_n = \mathscr{D}_n \cup \mathscr{D}_n'$. Let f' be the extension of f to this new space according to equation (5). We would like to apply the theorem in the special case of the previous paragraph, but f' still has fixed points at $P_1, \ldots, P_n$ plus additional fixed points at $P_1', \ldots, P_n'$. Let t be the transformation that maps every point Q on one of the spheres $\mathscr{C}_1, \ldots, \mathscr{C}_n$ to its reflection Q' across the equator and vice versa (see Figure 36.2), but otherwise leaves every point of $\mathscr{S}'$ fixed. t is continuous, so we can consider the composition $f'' = tf'$, which has *no* fixed points. One easily verifies that

$$\text{tr}(H_0(f'')) = \text{tr}(H_0(f))$$

and

$$\text{tr}(H_1(f'')) = \text{tr}(H_1(f))$$

However, because t reflects the spheres $\mathscr{C}_1, \ldots, \mathscr{C}_n$, thereby reversing their orientations, instead of (7) we have

$$\text{tr}(H_2(f'')) = \text{tr}(H_2(f)) - I(P_1) - I(P_2) - \cdots - I(P_n)$$

Since f'' has *no* fixed points, we obtain

$$\begin{aligned} 0 &= \text{tr}(H_0(f'')) - \text{tr}(H_1(f'')) + \text{tr}(H_2(f'')) \\ &= \text{tr}(h_0) - \text{tr}(h_1) + \text{tr}(h_2) - I(P_1) - I(P_2) - \cdots - I(P_n) \end{aligned}$$

which proves the Lefschetz fixed point formula.

Exercises

9. Supply the compactness argument required to complete the proof of the Lefschetz fixed point theorem.

10. Show that $L(f) = 1$ for all continuous transformations of the disk. Deduce the Brouwer fixed point theorem from the Lefschetz fixed point theorem.

11. Show that $L(f) = 1$ for all transformations of the projective plane. State and prove a fixed point theorem for the projective plane using this result. (Find another proof of this fixed point theorem by applying the Brouwer fixed point theorem directly.)

12. Find the Lefschetz number of the identity transformation on a surface $\mathscr{S}$.

13. Let f be a continuous transformation of the sphere to itself. Show that $L(f) = b + 1$, where b is the Brouwer degree of f. It follows that if f is fixed point free, then $b = -1$. This is the most important theorem on fixed points for the sphere.

14. Prove that a transformation of the sphere that maps the sphere into a proper subset of itself has a fixed point.

15. Given two transformations of the sphere f and g, prove that either f or g or $f \circ g$ has a fixed point.

16. If f is a continuous transformation of a sphere, prove that either f or f^2 has a fixed point.

17. If f is a continuous transformation of a sphere, prove that either f has a fixed point or f interchanges two diametrically opposite points.

18. Show that the transformation f_B of the torus defined in §35 always has a fixed point. Show that the sum of the indexes of these fixed points is $\det[I - B]$. Find the fixed points of the transformations corresponding to the following matrixes:

(a) $\begin{bmatrix} 4 & 1 \\ 1 & 2 \end{bmatrix}$ (b) $\begin{bmatrix} -1 & 0 \\ 1 & 2 \end{bmatrix}$ (c) $\begin{bmatrix} -2 & -1 \\ 0 & -1 \end{bmatrix}$

(d) $\begin{bmatrix} 2 & 3 \\ 5 & 1 \end{bmatrix}$ (e) $\begin{bmatrix} 6 & 3 \\ 3 & 2 \end{bmatrix}$ (f) $\begin{bmatrix} 2 & 1 \\ 0 & 1 \end{bmatrix}$

19. Find a fixed point free transformation of the torus.

---

§37 HOMOTOPY

No introduction to topology is complete without some mention of homotopy, the other great branch of algebraic topology. To obtain an intuitive picture of homotopy, consider two topological transformations f and g from a given path γ into the plane $\mathscr{P}$ (Figure 37.1). The image of γ under these transformations is a pair of paths γ_f and γ_g in $\mathscr{P}$. Imagine these two curves as rubber bands stretched into two positions in the plane. Intuitively we say that γ_f can be **deformed** into γ_g if by stretching and pulling the rubber band γ_f can be continuously moved in the plane from its original position until it coincides

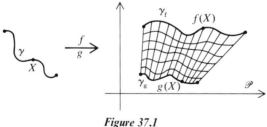

Figure 37.1

with the rubber band γ_g. In this movement the rubber must never be broken, so that it always forms a continuous path. Figure 37.1 illustrates this with several intermediate paths between γ_f and γ_g. At the same time, each point $f(X)$ on γ_f must follow a continuous path to the corresponding point $g(X)$ on γ_g. This situation is described by saying the two transformations f and g

are **homotopic**. Thus homotopy is intended as a precise mathematical description of the intuitive idea of continuous deformation. Here is the technical definition.

Definition

Let f, g be two continuous transformations from a topological space $\mathscr{S}$ to a space $\mathscr{T}$. Then f and g are called **homotopic** *if there is a family p_t of transformations from $\mathscr{S}$ to $\mathscr{T}$, one for each number $t \in [0, 1]$ such that*

(a) $p_0 = f$.

(b) $p_1 = g$.

(c) $p_t(X)$ *is* **jointly continuous** *in the two variables $t \in [0, 1]$ and $X \in \mathscr{S}$.*

This definition simply makes explicit the idea that the transformation f can be deformed into g. The intermediate stages of the deformation are represented by the transformations p_t for $0 < t < 1$. The variable t may be thought of as time. The transformations p_t then describe the motion of the transformation f into the transformation g. The key idea in this definition is that of joint continuity. This condition ensures among other things that the transformations $t \to p_t(X)$ are continuous for fixed X.

Definition

Let $f(X, Y)$ be a function of two variables, where X belongs to one topological space $\mathscr{S}_1$, Y belongs to another space $\mathscr{S}_2$, and $f(X, Y)$ belongs to a third space $\mathscr{S}_3$. Let $A = \{X_\sigma\}$ be a subset of $\mathscr{S}_1$ and $B = \{Y_\sigma\}$ be a matching subset of $\mathscr{S}_2$. Let $C = \{f(X_\sigma, Y_\sigma)\}$ be the corresponding subset of $\mathscr{S}_3$. The function f is called **jointly continuous** *if whenever X is near A and Y is near B, then $f(X, Y)$ is near C.*

An example of homotopic transformations is provided by a rotation of the sphere. Here the operation of rotating is a continuous deformation of the rotation itself into the identity transformation. Thus all rotations are homotopic to the identity.

Definition

A transformation of a space $\mathscr{S}$ that is homotopic to the identity is called a **deformation** *of $\mathscr{S}$.*

An example of a transformation that is not a deformation is the transformation of the sphere that interchanges diametrically opposite points. It is not obvious that this transformation is not a deformation, but this is one of the many consequences of the following theorem, the main theorem connecting homotopy with homology.

Theorem

Let f and g be homotopic transformations from a compact space $\mathscr{S}$ to a space $\mathscr{T}$. Then $H_k(f) = H_k(g)$, $k = 0, 1, 2$.

The idea of the proof is to divide the unit interval into subintervals $0 = t_0 < t_1 < t_2 < \cdots < t_n = 1$ in such a way that in the corresponding sequence of transformations $p_{t_0}, p_{t_1}, \ldots, p_{t_n}$, adjacent transformations are such close approximations of each other that they have a common simplicial approximation, and hence induce the same homology homomorphisms.

Let $\mathscr{L}$ be a triangulation of $\mathscr{T}$. We claim that there exists a $\delta > 0$ and a triangulation $\mathscr{K}$ of $\mathscr{S}$ so fine that for any vertex P of $\mathscr{K}$ and pair of numbers s, t from the unit interval such that $|s - t| < \delta$ there exists a vertex R of $\mathscr{L}$ such that both

$$p_t(\mathrm{st}(P)) \subseteq \mathrm{st}(R) \quad \text{and} \quad p_s(\mathrm{st}(P)) \subseteq \mathrm{st}(R) \tag{1}$$

Once this is established, the theorem is proven: according to (1), by defining $\tau(P) = R$, we obtain a simplicial transformation that is a simplicial approximation of p_s and p_t simultaneously. Thus (1) implies that the transformations p_s and p_t have the same homology homomorphisms whenever $|s - t| < \delta$. The proof is completed by choosing the partition $0 = t_0 < t_1 < \cdots < t_n = 1$ so fine that the distance between adjacent numbers is less than δ.

The proof of (1) involves a compactness argument of the usual sort. For a change we give the details. Suppose, contrary to what we want to prove, that for every $\delta > 0$ and every triangulation $\mathscr{K}$ of $\mathscr{S}$ there is a pair of numbers s, t with $|s - t| < \delta$ and a vertex P of $\mathscr{K}$ such that $p_s(\mathrm{st}(P))$ and $p_t(\mathrm{st}(P))$ are *not* both contained in $\mathrm{st}(R)$ for any vertex R of $\mathscr{L}$. Let $\mathscr{K}$ be any triangulation of $\mathscr{S}$, and let $\{\mathscr{K}_n\}$ be the sequence of barycentric subdivisions of $\mathscr{K}$. Then for each $\mathscr{K}_n$ there is a vertex P_n of $\mathscr{K}_n$ and a pair of numbers s_n, t_n with $|s_n - t_n| < 1/n$ such that $p_{s_n}(\mathrm{st}(P_n))$ and $p_{t_n}(\mathrm{st}(P_n))$ are not both contained in $\mathrm{st}(R)$ for any vertex R of $\mathscr{L}$. By compactness of $\mathscr{S}$ and compactness of the unit interval, there is a point P of $\mathscr{S}$ near the sequence $\{P_n\}$ and a number s in the unit interval near the sequence $\{s_n\}$. Clearly s is also near the sequence $\{t_n\}$. By joint continuity, the point $p_s(P)$ is near both sequences $\{p_{s_n}(P_n)\}$ and $\{p_{t_n}(P_n)\}$. Now, $p_s(P)$ belongs to $\mathrm{st}(R)$ for *some* vertex R of $\mathscr{L}$. By construction, for each n there exists a point Q_n of $\mathrm{st}(P_n)$ such that either $p_{s_n}(Q_n)$ or $p_{t_n}(Q_n)$ is *not* in $\mathrm{st}(R)$. One of these cases (p_{s_n} or p_{t_n}) must occur infinitely often, so that by restricting our attention to a subsequence, we can suppose, without loss of generality, that always it is $p_{s_n}(Q_n)$ that is not in $\mathrm{st}(R)$. However, since the triangulations $\mathscr{K}_n$ grow finer and finer, it is clear that P is near the sequence $\{Q_n\}$. Therefore, by joint continuity, $p_s(P)$ is near the sequence $\{p_{s_n}(Q_n)\}$. This contradicts the fact that no points of this sequence are in the neighborhood $\mathrm{st}(R)$ of $p_s(P)$, completing the proof of the theorem.

Exercises

1. Let f, g, h be continuous transformations. Show that homotopy has the following algebraic properties:

   (a) If f is homotopic to g, then g is homotopic to f.

   (b) If f is homotopic to g and g is homotopic to h, then f is homotopic to h.

2. A **constant** transformation $f:\mathscr{S} \to \mathscr{S}$ is a transformation taking the same value for every point $P \in \mathscr{S}$. Compute the homomorphisms $H_k(f)$ for a constant transformation.

3. A space $\mathscr{S}$ is **contractible** if the identity transformation $I:\mathscr{S} \to \mathscr{S}$ is homotopic to a constant transformation. Show that contractible spaces are connected. Show that the disk is contractible.

4. Prove that a contractible space $\mathscr{S}$ has trivial first and second homology groups.

5. Show that the Brouwer degree of a deformation is $+1$. Describe the homology transformations of deformations and find their Lefschetz number.

6. Show that the transformation of the sphere that interchanges diametrically opposite points is not a deformation.

7. Prove that the rotations of the torus r_A are deformations. In contrast show that no two of the transformations f_B of the torus are homotopic.

8. Which surfaces have fixed point free deformations?

9. Prove that all continuous transformations of the disk are deformations.

TOPOLOGICAL DYNAMICS

Consider a vector field V in the plane. As explained in Chapter Two, the study of V is equivalent to the study of a system of differential equations

$$\frac{dx}{dt} = F(x, y)$$

$$\frac{dy}{dt} = G(x, y) \tag{1}$$

The solution paths of these equations, given parametrically by $t \to (x(t), y(t))$, form the phase portrait of the vector field V. Intuitively these paths are the paths that would be followed by a ball if it were placed in the plane and rolled about by forces directed by the vector field V. The topological analysis of these phase portraits has been a constantly recurring problem ever since Chapter Two. Here we discuss still another aspect. Let $x(t)$ and $y(t)$ be the solution of the system (1) with initial conditions $x(0) = x_0$, $y(0) = y_0$. In other words, $x(t)$, $y(t)$ are the parametric equations of the path that passes through a given point (x_0, y_0) at $t = 0$. Fix a time t; then the transformation $f_t(x_0, y_0) = (x(t), y(t))$ is a continuous transformation of the plane. Intuitively, f_t is the transformation that results if the plane were covered by marbles that were all permitted to roll subject to the force field V for exactly t minutes. Clearly the family of transformations $\{f_t\}$ contains within it complete information regarding the phase portrait of V. These transformations represent another means of attacking the vector field topologically. The basic properties of the family $\{f_t\}$ are summarized in the following definition.

Definition

Let $\mathscr{S}$ be a topological space. Let $\{f_t\}$ be a family of continuous transformations of $\mathscr{S}$, one for each real number t. Suppose in addition that

(a) $f_0(P) = P$ *for* $P \in \mathscr{S}$.

(b) $f_t(f_s(P)) = f_{t+s}(P)$.

(c) $f_t(P)$ *is jointly continuous in t and P.*

Then $\{f_t\}$ is called a **flow on** $\mathscr{S}$.

The idea of a flow on a topological space $\mathscr{S}$ extends far beyond the flows in the plane arising from vector fields. Flows not only arise on other surfaces (from tangent vector fields), but occur also on nonorientable surfaces, complexes of higher dimension than two, and also on all sorts of abstract topological spaces. Flows are a vast generalization of vector fields. The study of flows is called **topological dynamics** and is an abstract form of the qualitative theory of differential equations.

Verification of the first two properties of a flow is usually trivial. The first (a) simply states that the transformation f_0 is the identity; while the second (b), called the **group property**, asserts that the family $\{f_t\}$ forms a group isomorphic to the group of real numbers. In the case of a flow arising from a vector field V, (a) is true by definition, while (b) simply says that if you let a marble roll for s minutes and then for another t minutes, the result is the same as a roll of $s + t$ minutes. The last condition (c) is the most technical, but naturally the most important for topology. For vector fields, joint continuity of $f_t(P)$ follows from the fundamental existence and uniqueness theorems on systems of differential equations (see, for example, Hirsch and Smale [13]).

Let $\{f_t\}$ be a flow on a space $\mathscr{S}$. Given a point P, the set of points $\{f_t(P)\}$ forms a path in $\mathscr{S}$ passing through P. The collection of all paths formed in this way is the **phase portrait** of the flow. There is exactly one such path passing through each point of $\mathscr{S}$ except the points P fixed by the transformations f_t. These fixed points are called the **critical** points of the flow.

For example, consider the vector field $V(x, y) = (x, 2y)$. In order to find parametric equations for the paths of the corresponding flow, we must solve the differential equations

$$\frac{dx}{dt} = x, \qquad x(0) = x_0$$

$$\frac{dy}{dt} = 2y, \qquad y(0) = y_0$$

we obtain $x = e^t x_0$, $y = e^{2t} y_0$. The flow $\{\phi_t\}$, where $\phi_t(x, y) = (e^t x, e^{2t} y)$, has the phase portrait shown in Figure 7.2.

Let $\{f_t\}$ be a flow. Then all the transformations f_t are homotopic to the identity f_0; hence all are deformations. This leads to the following generalization of the Poincaré index theorem.

Theorem

The sum of the indexes of the critical points of a flow on a surface $\mathscr{S}$ is the Euler characteristic of $\mathscr{S}$.

Exercises

10. Find explicit formulas for the flows in the plane corresponding to the following vector fields, and sketch the phase portraits:

(a) $V(x, y) = (-x, y)$ (b) $V(x, y) = (x, xy^2)$

(c) $V(x, y) = (x^2, y^2)$ (d) $V(x, y) = (x + y, x - y)$

11. Verify that the following are flows on the complex sphere, sketch the phase portrait, and verify the conclusion of the Poincaré index theorem:

(a) $f_t(z) = \dfrac{\cos(t)z + \sin(t)}{-\sin(t)z + \cos(t)}$ (b) $f_t(z) = \dfrac{(t + 1)z + t}{tz + (1 - t)}$

(c) $f_t(z) = \dfrac{\cosh(t)z + \sinh(t)}{\sinh(t)z + \cosh(t)}$ (d) $f_t(z) = z + t$

12. Let $\mathscr{S}$ be the cylinder in space with the equation $x^2 + y^2 = 1$, $0 \leq z \leq 1$. Sketch $\mathscr{S}$ together with the phase portrait of the flow given by

$$f_t(x, y, z) = (x \cos(t) - y \sin(t), x \sin(t) + y \cos(t), ze^{-t})$$

13. Let $A = (a, b)$ be a vector in the plane. Sketch the phase portrait of the flow $\{f_t\}$ on the torus, where $f_t = r_{tA}$.

Homotopy, the Brouwer degree, and the Lefschetz fixed point theorem all combine to prove the following important theorem.

The Fundamental Theorem of Algebra

Every polynomial equation, $z^n + a_{n-1}z^{n-1} + \cdots + a_1z + a_0 = 0$, *has at least one root in the complex plane for* $n \geq 1$.

Let $p(z) = z^n + a_{n-1}z^{n-1} + \cdots + a_0$. Since the conclusion of the theorem is clear if $n = 1$, we can assume that $n \geq 2$. The roots of the equation $p(z) = 0$ are the same as the fixed points of the continuous function $p(z) + z$. Therefore it suffices to prove that every polynomial of degree greater than or equal to two has a fixed point in the complex plane. Regarding p as a transformation of the sphere, the sum of the indexes of its fixed points is $b + 1$, where b is the Brouwer degree of p. Now p is homotopic to the polynomial $p_0(z) = z^n$ via the homotopy $p_t(z) = z^n + t(a_{n-1}z^{n-1} + \cdots + a_0)$. Therefore p and p_0 have the same Brouwer degree, so that $b = n$. It will follow that p has a fixed point in the plane just as soon as we show that the index of the fixed point at ∞ is not equal to $n + 1$.

To compute the index of p at ∞, let $t(z) = 1/z$, and consider the composition $f = t \circ p \circ t$. This moves the fixed point at ∞ to zero. We find that

$$f(z) = \frac{z^n}{a_0z^n + a_1z^{n-1} + \cdots + a_{n-1}z + 1}$$

In a suitably small neighborhood of zero, the denominator is nearly 1; therefore we expect that the index of f at zero will be the same as the index of z^n at zero, namely 1. To prove this, let $q(z) = a_0z^n + a_1z^{n-1} + \cdots + 1$, and choose the real number R so small that for $\|z\| \leq R$, $\|q - 1\| \leq \frac{1}{2}$. It follows that also $\|1/q\| \leq 2$. Therefore if, in addition, R is chosen so that $R \leq \frac{1}{4}$, then for $\|z\| = R$,

$$\|f(z)\| \leq 2\|z^n\| = 2R^n \leq \frac{R}{2}$$

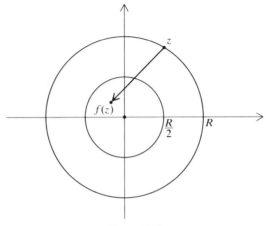

Figure 37.2

because $n \geq 2$. Thus as z travels around the circle $\|z\| = R$ of radius R, $f(z)$ is trapped inside the smaller circle of radius $R/2$ (Figure 37.2). Therefore the index of f at zero is one.

The fundamental theorem is one of the most important facts about the complex numbers. The first proof was given by Gauss in 1799. His treatment established the validity of operations with complex numbers that had formerly been viewed with suspicion by most mathematicians. Our proof resembles Gauss's in two important ways. First, both proofs are geometrical rather than algebraic, and second, both are *pure* existence proofs; that is, they prove the existence of a root but provide no means whatsoever for finding a root. Gauss's proof was the first pure existence proof. Before Gauss, going back to the Greeks, who invented the idea of proof, existence proofs were always constructive; that is, the object, whose existence was being proven, was actually found or constructed from the material at hand. Since Gauss, pure existence proofs have played an increasingly important role in mathematics, not without controversy. There is something about proving the existence of something, without being able to produce that thing, that has made many mathematicians uneasy. As Gordan said of a later pure existence proof, "This is not mathematics, it is theology." In the late nineteenth century, a group of mathematicians began to deny the validity of all but constructive proofs. These constructivists remain to this day a small but important group. Some of our earlier proofs have also been nonconstructive, notably the proof of Brouwer's fixed point theorem. Brouwer himself became a constructivist, one of the most prominant of this century, and repudiated some of his early results. Today the majority of mathematicians would agree with the constructivists to the extent of preferring a constructive proof over any other, but would not reject pure existence proofs, which are more important now than ever before. Incidentally, Gauss, who many feel was the greatest of all mathematicians, eventually gave four more proofs of the fundamental theorem of algebra.

Exercises

14. Prove that a polynomial takes on every complex value at least once.

15. Let $f(z)$ be a rational function considered as a transformation of the complex sphere. If at a point z_0 the value of f is $w_0 = f(z_0)$, we define the **weight** with which f takes this value as the index of z_0 as a fixed point of the transformation $f(z) - w_0 + z$. Prove that for every fixed w_0 if there are only a finite number of points z_0 such that $f(z_0) = w_0$, then the sum of the weights of these points is the Brouwer degree of f.

---

§38 OTHER HOMOLOGIES

The homology discussed in this book is called simplicial homology. You cannot help but be aware of the very special nature of this theory. Almost every theorem we have proved is a special case of a much more general theorem. With the same techniques and a somewhat more cumbersome notation, simplicial homology extends to three-dimensional complexes, four-dimensional complexes, and still higher; while instead of the integers we can use rational numbers, real numbers, or even complex numbers as coefficients. All of these generalizations would be based, however, on the same geometric ideas that we have chosen to introduce in two dimensions. More interesting is the fact that there are other types of homology based on other geometric ideas. In this section we give a brief account of one of these.

The development of simplicial homology was paralleled by several rivals that grew in response to a number of problems. In the first place, simplicial homology applies only to complexes: topological spaces that are obtained by gluing together points, intervals, disks, cubes, etc.—all pieces of Euclidean spaces. The problem was to find a homology theory extending beyond spaces constructed out of Euclidean geometry. A second problem is the invariance theorem. Our proof of invariance has been based for surfaces on the classification theorem and for other two-dimensional complexes on the Hauptvermutung. In higher dimensions, neither of these is available: no classification theorem is known, while the Hauptvermutung remains a conjecture. The proof of invariance in these cases depends on a complicated analysis using simplicial approximation. This was achieved by Alexander in 1926 using singular homology. The rival homology theories not only brought in new geometric ideas but also greatly simplified the problem of invariance.

Definition

An **algebraic sequence** *is a sequence of groups and homomorphisms as follows:*

$$\mathscr{C}_0 \xleftarrow{\partial} \mathscr{C}_1 \xleftarrow{\partial} \mathscr{C}_2 \xleftarrow{\partial} \mathscr{C}_3 \xleftarrow{\partial} \mathscr{C}_4 \xleftarrow{\partial}$$

Such a sequence can be of any length, although here we need only consider sequences with three groups and two homomorphisms ∂. The sequence is called **homological** *if $\partial^2 = 0$. Elements in the range and kernel of ∂ are called, respectively,* **boundaries** *and* **cycles**, *with the convention that all the elements of $\mathscr{C}_0$ are called 0-cycles. In a homological sequence, every boundary is a cycle. The quotient groups of cycles by boundaries are called the* **homology groups** *of the sequence.*

This definition summarizes all the algebraic steps involved in the construction of homology groups. In order now to set up a homology theory, all that is required is the description of the particular homological algebraic sequence involved.

Exercise

1. Prove that every boundary is a cycle in a homological sequence.

SINGULAR HOMOLOGY

This is the most general homology theory, because it can be defined for any topological space $\mathscr{S}$. In general this requires an infinite sequence of homology groups. We will be content to describe the first three groups $H_k(\mathscr{S})$, $k = 0, 1, 2$.

To begin, we set up a **standard** k-simplex ($k = 0, 1, 2$) in the plane. The standard 0-simplex S_0 will be the origin, the standard 1-simplex S_1 will be

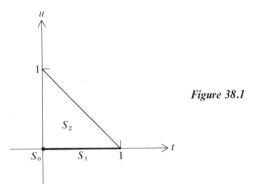

Figure 38.1

the unit interval on the x-axis, and the standard 2-simplex S_2 will be the triangle with vertexes $(0, 0)$, $(1, 0)$, and $(0, 1)$. The standard simplexes are shown in Figure 38.1.

Definition

A **singular k-simplex** *in $\mathcal{S}$ is a continuous function $s: S_k \to \mathcal{S}$ from the standard k-simplex S_k into $\mathcal{S}$. Thus the 0-simplexes are all points of $\mathcal{S}$, the 1-simplexes are all continuous paths in $\mathcal{S}$, and so forth. A* **singular k-chain** *is any sum*

$$a_1 C_1 + a_2 C_2 + \cdots + a_n C_n$$

where $C_1, C_2, \ldots, C_n$ are singular k-simplexes and $a_1, a_2, \ldots, a_n$ are any integers. The sum of two such chains is defined as the chain whose coefficients are the sums of the coefficients of the individual summands, just as in the integral simplicial homology (§29). The set of all these chains forms the group $\mathcal{C}_k$ of singular k-chains.

This definition reveals at once the power and the limitations of singular homology. On the one hand, the invariance of singular homology is almost trivial. No choices are made in setting up the singular chain groups. Every simplex (vertex, edge, or triangle) that can possibly be continuously embedded in $\mathcal{S}$ is thrown in as a singular simplex. Furthermore, every possible linear combination of such simplexes is considered a chain whether the simplexes can be fitted into a complex or not. Since the singular chain groups do not depend on the choice of a particular complex, there is nothing of which they need to be independent; their definition depends only on the topological space $\mathcal{S}$ itself. On the other hand, it follows that these groups are enormous. The actual computation of a singular chain group is a complete impossibility.

Thus the singular theory solves the problem of invariance at the expense of introducing a new problem: How are the groups $\mathscr{C}_k$ to be computed? We will see the modern resolution of this problem at the end of this section, but first let us define the singular boundary operator and thus complete the definition of singular homology.

The boundary of a singular simplex is defined with the aid of some special transformations among the standard simplexes. These are defined in Figure 38.2. Observe that the transformations d_1, d_2 carry S_0 to the two endpoints

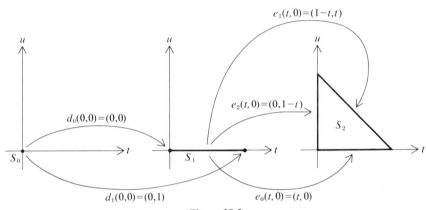

Figure 38.2

of S_1, while e_1, e_2, e_3 carry S_1 to each of the three sides of the triangle S_2. Now, the boundary of a singular simplex is defined as follows: if $s:S_1 \to \mathscr{S}$ is a 1-simplex, ∂s is the 0-chain

$$\partial s = sd_1 - sd_0$$

and if $s:S_2 \to \mathscr{S}$ is a 2-simplex, ∂s is the 1-chain

$$\partial s = se_2 + se_1 + se_0$$

The boundary of a chain is determined from the boundary of its simplexes by additivity. This completes the description of singular homology.

Exercises

2. Prove that the algebraic sequence of singular chain groups and boundary operators is homological.

3. Let f be a continuous transformation from a topological space $\mathscr{S}$ to the space $\mathscr{T}$. Let $s:S_k \to \mathscr{S}$ be a k-simplex on $\mathscr{S}$. The composition $fs:S_k \to \mathscr{T}$ is a k-simplex on $\mathscr{T}$. Setting $f_k(s) = fs$ and extending by additivity, we obtain a homomorphism f_k from the group $\mathscr{C}_k(\mathscr{S})$ of k-chains on $\mathscr{S}$ to the group $\mathscr{C}_k(\mathscr{T})$ of k-chains on $\mathscr{T}$. Prove that the following diagram commutes:

$$
\begin{array}{ccccc}
\mathscr{C}_0(\mathscr{S}) & \xleftarrow{\ \partial\ } & \mathscr{C}_1(\mathscr{S}) & \xleftarrow{\ \partial\ } & \mathscr{C}_2(\mathscr{S}) \\
\downarrow{\scriptstyle f_0} & & \downarrow{\scriptstyle f_1} & & \downarrow{\scriptstyle f_2} \\
\mathscr{C}_0(\mathscr{T}) & \xleftarrow[\ \partial\ ]{} & \mathscr{C}_1(\mathscr{T}) & \xleftarrow[\ \partial\ ]{} & \mathscr{C}_2(\mathscr{T})
\end{array}
$$

4. Let $f:\mathscr{S} \to \mathscr{T}$ be a continuous transformation. Using the previous exercise, explain how f induces a homomorphism h_k from the kth singular homology group of $\mathscr{S}$ to the kth singular homology group of $\mathscr{T}$.

THE EILENBERG-STEENROD AXIOMS

There are still other homology theories, for example, the Cech theory and the Vietoris theory. Each has its peculiar advantages and disadvantages. Despite their differences, all of these theories are obviously related in some way. They are all clearly about the same thing. The precise nature of their relationship was for many years obscure, but in 1952 Eilenberg and Steenrod [7] succeeded in formulating *axioms* for homology that made explicit the crucial properties these theories have in common. We cannot describe all of these axioms here, because several involve properties of homology we have not studied. However the general drift of the axioms can be explained as follows. In the first place, a homology theory must be defined for a sufficiently large class of spaces—for all triangulable spaces, for example. Then for each space $\mathscr{S}$, the homology theory has to supply a series of groups $H_0(\mathscr{S})$, $H_1(\mathscr{S})$, $H_2(\mathscr{S}), \ldots$. The series is infinite, but as a practical matter for the spaces most commonly studied the groups are all trivial from some point on. Finally, for every continuous transformation $f:\mathscr{S} \to \mathscr{T}$, the homology theory must supply a series $H_0(f)$, $H_1(f)$, $H_2(f), \ldots$ of induced group homomorphisms $H_k(f):H_k(\mathscr{S}) \to H_k(\mathscr{T})$. The homology groups and homomorphisms must satisfy the following axioms:

1. If f is the identity transformation, then $H_k(f)$ is the identity homomorphism.

2. If $f:\mathscr{S} \to \mathscr{T}$ and $g:\mathscr{T} \to \mathscr{U}$ are both continuous transformations, then $H_k(fg) = H_k(f)H_k(g)$.

3. If f and g are homotopic, then $H_k(f) = H_k(g)$.

4. If the space $\mathscr{S}$ is a single point, then $H_k(\mathscr{S})$ is the trivial group for all k.

There are three further axioms, but those given above convey the spirit of the axioms as a whole. Note that the emphasis is on the transformations f rather than the spaces $\mathscr{S}$. It is characteristic of modern homology theory that the transformations are regarded as the important objects of study. The realization of the importance of transformations was a gradual movement spread over the first half of the present century. Appropriately in this book, written on quasi-historical lines, these axioms were not encountered (as theorems of simplicial homology) until this last chapter.

The most wonderful property of the Eilenberg-Steenrod axioms is that for triangulable spaces they are *categorical*, meaning that Eilenberg and Steenrod proved that *all homology theories satisfying their axioms must give the same results both for spaces and transformations for all triangulable spaces*. As a consequence, each homology theory can be used where its peculiar nature makes it most useful, and the results become immediately applicable to the other theories.

Exercise

5. Prove that singular homology satisfies the four Eilenberg-Steenrod axioms given above.

Notes. The treatise by Eilenberg and Steenrod [7] is a classic, but an accessible classic. Much of this chapter is based on the treatment there of continuous transformations. The books on homology cited at the end of Chapter Five also contain material on continuous transformations. For homotopy, you might well start with Greenberg [9] and then consult the books suggested there. Sansone and Gerretsen [26] give an excellent introduction to complex analysis; Volume II is particularly concerned with geometry. Its discussion of covering spaces was used in preparing this book. A standard reference on topological dynamics is the appendix to Nemytskii and Stepanov [24]. This book is especially valuable because it discusses the qualitative theory of differential equations in both concrete and abstract terms.

supplement

Topics in Point Set Topology

§39 CRYPTOMORPHIC VERSIONS OF TOPOLOGY

While lacking the geometric appeal and concrete applications of combinatorial topology, point set topology has a unique importance. Next to set theory it is the most fundamental branch of modern mathematics. The concepts defined in point set topology provide the language in which ideas are expressed in nearly all branches of mathematics. Point set topology is therefore one of those subjects that must be studied by all mathematicians. The object of this chapter is to introduce you to a significant portion of this language.

While in combinatorial topology both the language of the subject and the machinery of proof are crucial to applications, in point set topology knowledge of the language alone is often sufficient. Thus in this supplement we will pay much less attention to proofs as compared with the body of the text. Rather than begin at the beginning, we can assume that you are familiar with §4 and start where that section left off. For reference, we restate the definition of a topological space.

Definition

A topological space is a set $\mathcal{S}$ together with a collection **N** *of subsets of $\mathcal{S}$ called* **neighborhoods** *satisfying the following axioms:*

(N_1) *Every point of $\mathcal{S}$ is in at least one neighborhood.*

(N_2) *If A and B are neighborhoods and P is a point of $A \cap B$, then there is a neighborhood C such that $P \in C \subseteq A \cap B$.*

In this approach to topology, the neighborhoods are used to define the concept of nearness and all the other notions of point set topology: open sets, closed sets, and so forth (see § 4). It is possible, however, to take one of these other ideas as the basis for the definition of topology. This leads to a cryptomorphic version of topology: the same subject is simply being studied in disguise. For example, open sets may be used as the fundamental notion of topology. The following theorem summarizes the necessary crucial properties.

Theorem 1

The collection **O** *of open sets in a topological space $\mathscr{S}$ possesses the following properties:*

(O_1) *The whole space $\mathscr{S}$ and the empty set are in* **O**.

(O_2) *The union of any number of sets from* **O** *is in* **O**.

(O_3) *The intersection of any finite number of sets from* **O** *is in* **O**.

Recall that a subset A of $\mathscr{S}$ is open if every point of A is not near the complement A' of A. Since no point is near the empty set and every point is near the whole space, (O_1) is obvious. Now consider a collection $\{A_\sigma\}$ of open sets, where the subscript σ can range over any set, finite or infinite. Let P be a point of the union $A = \cup_\sigma A_\sigma$. Then P is a member of some particular set A_j. Since P is not near A_j', and A_j' contains A', it follows that P is not near A'; hence A is open. This proves (O_2). To prove (O_3), it suffices to consider the intersection of just two open sets, say A_1 and A_2. Let $P \in A_1 \cap A_2$. Then P is not near A_1' or A_2'. Thus there exist neighborhoods B_1 and B_2 of P such that $B_1 \subseteq A_1$ and $B_2 \subseteq A_2$. By axiom (N_2), P has a neighborhood $B \subseteq B_1 \cap B_2 \subseteq A_1 \cap A_2$. Thus P is not near $(A_1 \cap A_2)'$. This proves (O_3).

The way that open sets offer an alternative development of topology is described in the following theorem.

Theorem 2

Let $\mathscr{S}$ be a set together with a collection **O** *of subsets satisfying the axioms $(O_1), (O_2),$ and (O_3). Then* **O** *can be used as the collection of neighborhoods for a topology on $\mathscr{S}$ whose open sets are exactly the sets of* **O**.

To prove Theorem 2, we must first show that the collection **O** satisfies the axioms (N_1) and (N_2). Actually this is not difficult, because (N_1) follows straight from (O_1), and (N_2) from (O_3). Thus **O** can be used as the set of neighborhoods for a topological space on the set $\mathscr{S}$. It remains to prove that the open sets of this topology are exactly the sets of **O**. Since a neighborhood is considered a neighborhood for each of its points, neighborhoods are always open sets. Thus the sets of **O** are open in $\mathscr{S}$. Conversely, let A be an open subset of $\mathscr{S}$. Then every point P of A has a neighborhood $B_P \in$ **O** that is entirely contained in A. Therefore the union $\cup_P B_P$ is A, so that by (O_2), A is in **O**.

Theorem 2 describes the way topological spaces are defined most often in practice. It is usual, in other words, to let the neighborhoods be all of the open sets. Since the closed sets are simply the complements of the open sets, it follows that the closed sets offer another cryptomorphic version of topology. This is described in the next two theorems, whose proofs are left as exercises.

Theorem 3

In a topological space $\mathscr{S}$, the collection **C** *of closed sets possesses the following properties:*

(C_1) *The whole space and the empty set are in* **C**.

(C_2) *The intersection of any number of sets from* **C** *is in* **C**.

(C_3) *The union of a finite number of sets in* **C** *is in* **C**.

Theorem 4

Let $\mathscr{S}$ be a set together with a collection of sets **C** *satisfying the axioms* (C_1), (C_2), *and* (C_3). *Then the collection of complementary sets can be used to define a topology on $\mathscr{S}$ whose closed sets are exactly the sets of* **C**.

There is no end to the cryptomorphic versions of topology. Still another example is furnished by the operation of closure. Let A be any subset of a space $\mathscr{S}$. Recall that the closure of A, written $\bar{A}$, is the set A together with all its near points. The closure is the smallest closed set containing A.

Theorem 5

The closure operation has the following properties:

(K_1) $\bar{\varnothing} = \varnothing$.

(K_2) $A \subseteq \bar{A}$.

(K_3) $\overline{A \cup B} = \bar{A} \cup \bar{B}$.

(K_4) $\bar{\bar{A}} = \bar{A}$.

Theorem 6

Let $\mathscr{S}$ be a set together with an operation $A \to \bar{A}$ defined for all subsets of $\mathscr{S}$ and satisfying the closure axioms (K_1), (K_2), (K_3), and (K_4). Then the collection $\mathbf{C}$ of sets A such that $A = \bar{A}$ satisfies the axioms for closed sets. Furthermore, in the resulting topology on $\mathscr{S}$, the given operation is the closure operation.

The proof of Theorem 5 is left as an exercise. Turning to the proof of Theorem 6, let $\mathbf{C}$ be the collection of sets such that $A = \bar{A}$. The fact that $\mathbf{C}$ satisfies the axiom (C_1) follows straight from (K_1) and (K_2), while (C_3) follows immediately from (K_3). To prove (C_2), let $\{A_\sigma\}$ be a collection of sets from $\mathbf{C}$ and let $A = \cap_\sigma A_\sigma$. For every σ, $A \subseteq A_\sigma$. Using (K_3), it follows that $\bar{A} \subseteq \bar{A}_\sigma = A_\sigma$; therefore $\bar{A} \subseteq A$. But since $A \subseteq \bar{A}$ by (K_2), it follows that $A = \bar{A}$, proving (C_2).

Thus according to Theorem 4, $\mathbf{C}$ defines a topology for $\mathscr{S}$. For a subset A of $\mathscr{S}$ we will write $\mathrm{cl}(A)$ for the closure of A in this topology until we prove that $\mathrm{cl}(A) = \bar{A}$. Now by (K_4), $\bar{A}$ is a set in $\mathbf{C}$; therefore $\bar{A}$ is a closed set in this topology, a closed set containing A, according to (K_2). Since $\mathrm{cl}(A)$ is the smallest closed set containing A, it follows that $\mathrm{cl}(A) \subseteq \bar{A}$. On the other hand, since $A \subseteq \mathrm{cl}(A)$, it follows using (K_3) that $\bar{A} \subseteq \overline{\mathrm{cl}(A)} = \mathrm{cl}(A)$. Thus $\bar{A} = \mathrm{cl}(A)$.

Exercises

1. Prove Theorems 3, 4, and 5.

2. Let $\mathscr{S}$ be a topological space with neighborhoods $\mathbf{N}$. Prove that every open set is a union of sets from $\mathbf{N}$.

3. A **base** for a topological space $\mathscr{S}$ is a collection of sets $\mathbf{B}$ such that every open set of $\mathscr{S}$ is a union of sets from $\mathbf{B}$. According to the previous exercise, $\mathbf{N}$ is a base for $\mathscr{S}$. Conversely, prove that every base $\mathbf{B}$ satisfies the neighborhood axioms (N_1) and (N_2).

4. Let $\mathbf{B}$ be a base for the space $\mathscr{S}$. Prove that if the sets in B are used as neighborhoods, then $\mathscr{S}$ would have the same open sets as before. Thus every base offers an alternative development of the topology of $\mathscr{S}$.

5. Verify that the following collections of sets are bases for the usual topology of the plane:

(a) all open rectangles

(b) all open squares

(c) all open ellipses whose major axis is twice the minor axis.

6. Let G be a group. For any subset A of G, let $sp(A)$ denote the set of all elements dependent upon A. Discover which of the closure axioms the operation sp satisfies.

7. Let $\mathscr{S}$ be a topological space. For any subset A, the interior of A, $I(A)$, is the set of all points P of A that are not near A'. The operation I satisfies axioms analogous to the closure axioms. Find them, and prove that an operation satisfying them provides another cryptomorphic version of topology.

8. Let the set $\mathscr{S}$ have two topologies whose open sets are $\mathbf{O}_1$ and $\mathbf{O}_2$. The topology $\mathbf{O}_1$ is called **stronger** than $\mathbf{O}_2$ if every set in $\mathbf{O}_2$ is also in $\mathbf{O}_1$. In this case $\mathbf{O}_2$ is called **weaker** than $\mathbf{O}_1$. What is the weakest topology possible on $\mathscr{S}$? What is the strongest?

9. Arrange a diagram comparing the following plane topologies: usual, discrete, indiscrete, sector, finite, and either/or. Note that two topologies can be **incomparable**, meaning that neither is stronger than the other.

CONTINUOUS TRANSFORMATIONS

For every cryptomorphic version of topology there is a corresponding description of continuous functions. Let us begin by reviewing the definition in §4.

Definition

A function $f:\mathscr{S} \to \mathscr{T}$ between topological spaces is **continuous** *if whenever P is near A in $\mathscr{S}$, then $f(P)$ is near $f(A)$ in $\mathscr{T}$.*

The most important cryptomorphic description of continuity is the one using open sets.

Theorem

A function $f:\mathscr{S} \to \mathscr{T}$ is continuous if and only if for every open set B in $\mathscr{T}$ the inverse image $f^{-1}(B) = \{P \mid f(P) \in B\}$ is open in $\mathscr{S}$.

To prove the theorem, suppose first that f is continuous, and let B be an open set in $\mathscr{T}$. To prove that $f^{-1}(B)$ is open in $\mathscr{S}$, let $P \in f^{-1}(B)$. By continuity, if P is near $f^{-1}(B)'$, then $f(P)$ is near $f(f^{-1}(B)') = B'$. This is impossible, because B is open; therefore P is *not* near $f^{-1}(B)'$ and $f^{-1}(B)$ is open. Conversely, suppose that f has the property that the inverse image of an open set is open. By Theorem 1 we can assume that the neighborhood systems of $\mathscr{S}$ and $\mathscr{T}$ include all open sets. Then the property we assume for f is that the inverse image of a neighborhood is a neighborhood. Let P be a point of $\mathscr{S}$ and A a subset of $\mathscr{S}$ such that $P \leftarrow A$. Let B be a neighborhood of $f(P)$. Then $f^{-1}(B)$ is a neighborhood of P. Because $P \leftarrow A$, $f^{-1}(B)$ contains a point Q of A. Therefore B contains the point $f(Q)$ of $f(A)$. This proves that $f(P) \leftarrow f(A)$, so f is continuous.

The characterization of continuity by open sets often appears backward at first, and there is a tendency to want to replace it with the following definition.

Definition

A transformation $f : \mathscr{S} \to \mathscr{T}$ *is called* **open** *if* $f(A)$ *is open in* $\mathscr{T}$ *whenever A is open in* $\mathscr{S}$.

However, continuous transformations are *not* usually open. For example, the function $f(x) = x^2$ is continuous but not open, since the image of the open interval $(-1, 1)$ is the half open–half closed interval $[0, 1)$. One simply must get used to the backward nature of continuity when using open sets.

Exercises

10. Let $f : \mathscr{S} \to \mathscr{T}$ be a continuous function. Prove that f has these properties:

(a) If A is closed in $\mathscr{T}$, then $f^{-1}(A)$ is closed in $\mathscr{S}$.

(b) For any set A of $\mathscr{T}$, $\overline{f^{-1}(A)} = f^{-1}(\overline{A})$.

Conversely, prove that each of these properties characterizes continuity.

11. Find a cryptomorphic version of continuity using the interior operation.

12. Let $f : \mathscr{S} \to \mathscr{T}$ be a transformation. Let **B** be a base for $\mathscr{T}$. Prove that if $f^{-1}(A)$ is open for each $A \in \mathbf{B}$, then f is continuous. Prove a similar result for open functions involving a base for $\mathscr{S}$.

METRIC SPACES

An important part of the story of point set topology has been the search for the crucial topological properties that characterize the usual spaces (the real line, the plane, and so forth) studied in analysis. Among the properties turned up in this search, none is more important than that of being a metric space.

Definition

A **metric space** *is a set* $\mathscr{S}$ *together with a* **distance function** $d(P, Q)$, *a function of two variables on* $\mathscr{S}$ *satisfying the following axioms:*

(D_1) $d(P, Q)$ *is zero if and only if* $P = Q$.

(D_2) $d(P, Q) = d(Q, P)$.

(D_3) $d(P, R) \le d(P, Q) + d(Q, R)$ (*triangle law*).

Given a point P in a metric space $\mathscr{S}$, the **disk about P of radius r** is defined by $N_r(P) = \{Q \,|\, d(P, Q) < r\}$. The set **N** of all disks in $\mathscr{S}$ satisfies the neighborhood axioms (N_1) and (N_2). Thus every metric space is a topological space. However, not every topological space is a metric space. Metric spaces are not a cryptomorphic version of topological spaces in general, but rather represent a special type of space. All the usual spaces are metric spaces, and some of the unusual ones as well. For example, a space with the discrete topology is always a metric space (Exercise 14).

Exercises

13. Prove that the disks **N** in a metric space satisfy the neighborhood axioms.

14. Let $\mathscr{S}$ be any set, and define $d(P, Q)$ as 1, unless $P = Q$ when $d(P, Q) = 0$. Show that this defines a distance function for which the resulting topology is the discrete topology.

15. Let A be a subset of a metric space $\mathscr{S}$. For a point P the distance from P to A is defined by $d(P, A) = \min_{Q \in A} d(P, Q)$. Prove that

$$\bar{A} = \{P \,|\, d(P, A) = 0\} \tag{1}$$

16. In a metric space $\mathscr{S}$ let the closure operation be defined by (1), and prove directly from the axioms (D_1), (D_2), and (D_3) that this operation satisfies the closure axioms.

17. Let the metric space $\mathscr{S}$ have distance function d. Show that $d' = d/(1 + d)$ is also a distance function that produces the same topology on $\mathscr{S}$. Thus it is possible to assume in a metric space that the distances are always less than one.

PRODUCT SPACES

Let $\mathscr{S}$ and $\mathscr{T}$ be topological spaces. The Cartesian product $\mathscr{S} \times \mathscr{T}$ is the set of all ordered pairs (P, Q), where $P \in \mathscr{S}$ and $Q \in \mathscr{T}$. As neighborhoods in $\mathscr{S} \times \mathscr{T}$ we use the **rectangle sets** $A \times B$, where A is a neighborhood in $\mathscr{S}$ and B is a neighborhood in $\mathscr{T}$. Then $\mathscr{S} \times \mathscr{T}$ becomes a topological space called the **product** of $\mathscr{S}$ and $\mathscr{T}$. The canonical example is the case where $\mathscr{S}$ and $\mathscr{T}$ are both the real line with the usual topology. Then $\mathscr{S} \times \mathscr{T}$ is the plane with the usual topology. The neighborhoods $A \times B$, being the products of intervals, are rectangles (Figure 39.1). This is the picture you should form in order to visualize all product spaces. The product is a very important construction in point set topology, providing a means for producing many new spaces. It is possible to define the Cartesian product for any number of spaces, even an infinite number. The latter case is particularly important, but there is no room for an adequate discussion here.

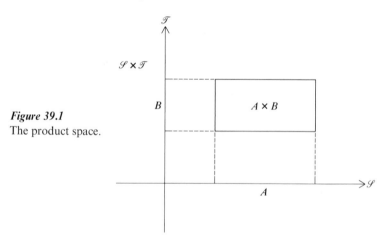

Figure 39.1
The product space.

With each topological property it is an interesting question whether the product $\mathscr{S} \times \mathscr{T}$ possesses the property if the two factors $\mathscr{S}$ and $\mathscr{T}$ possess the property. The following theorem is an example of the sort of result we have in mind.

Theorem

The product of metric spaces is a metric space.

If $\mathscr{S}$ has a metric d_1 and $\mathscr{T}$ has a metric d_2, then the sum d defined by

$$d((P_1, Q_1), (P_2, Q_2)) = d_1(P_1, P_2) + d_2(Q_1, Q_2) \qquad (2)$$

is a metric whose neighborhoods are easily shown to provide a base for the product topology as defined above (Exercise 20).

Exercises

18. Let $\mathscr{S} \times \mathscr{T}$ be a product space, and let $A \subseteq \mathscr{S}$ and $B \subseteq \mathscr{T}$. Prove that $(P, Q) \leftarrow (A \times B)$ if and only if $P \leftarrow A$ and $Q \leftarrow B$. Conclude that $\overline{A \times B} = \bar{A} \times \bar{B}$.

19. Let $\mathscr{S} \times \mathscr{T}$ be a product space. The two transformations $p_s : \mathscr{S} \times \mathscr{T} \to \mathscr{S}$ and $p_t : \mathscr{S} \times \mathscr{T} \to \mathscr{T}$ defined by $p_s(P, Q) = P$ and $p_t(P, Q) = Q$ are called **projections**. Prove that they are continuous and open.

20. Verify that the function d defined in (2) is a metric and that the topology that it defines on $\mathscr{S} \times \mathscr{T}$ is the product topology.

21. Let $f : \mathscr{S} \to \mathscr{T}$ be a continuous transformation. The subset $G_f = \{(P, f(P)) | P \in \mathscr{S}\}$ of the product space $\mathscr{S} \times \mathscr{T}$ is called the **graph of f**. Prove that G_f is a closed subset of $\mathscr{S} \times \mathscr{T}$ and is topologically equivalent to $\mathscr{S}$.

22. Prove that the product of connected spaces is connected.

23. Let $\mathscr{S}$ be the unit circle. Show that $\mathscr{S} \times \mathscr{S}$ is a torus.

---

§40 A BOUQUET OF TOPOLOGICAL PROPERTIES

The properties introduced here and in the next section, like the property of being a metric space, were all designed to isolate crucial features of the canonical examples of topological spaces.

SEPARATION PROPERTIES

These concern the separation of points and sets within a topological space.

Definition

Let $\mathscr{S}$ be a topological space, and let P and Q be any two distinct points of $\mathscr{S}$. $\mathscr{S}$ is called

(a) $\mathbf{T_0}$ if at least one of the points has a neighborhood excluding the other

(b) $\mathbf{T_1}$ if each point has a neighborhood excluding the other

*(c) **Hausdorff** (or $\mathbf{T_2}$) if there exists a pair of disjoint neighborhoods for P and Q*

This definition establishes a progression of stricter and stricter conditions on a space. Those that satisfy (c) will tend to be more like the usual spaces than those that merely satisfy (a). That these properties represent a genuine hierarchy among topological spaces can be demonstrated by examples: the indiscrete topology is not even T_0, the sector topology is T_0 but not T_1, the **modified either/or** topology (whose neighborhoods are either disks excluding the x-axis or infinite strips centered on the x-axis excluding finitely many points of the x-axis) is T_1 but not Hausdorff, and finally we have the following result.

Theorem A

Metric spaces are Hausdorff.

If P and Q are distinct points in a metric space, then $r = d(P, Q) \neq 0$. Thus $N_{r/3}(P)$ and $N_{r/3}(Q)$ are disjoint neighborhoods of P and Q.

Progressing along the separation hierarchy, properties associated with the usual topological spaces begin to appear. The following theorem is an example.

Theorem 1

In a T_1 space, points are closed sets.

At the next level, Hausdorff spaces have the important property that limits are unique. To state this properly requires a definition.

Definition

A sequence $\{P_n\}$ in a topological space $\mathscr{S}$ has **limit** P, written $\lim_n P_n = P$, if for every neighborhood A of P there is an integer N such that all the points P_N, P_{N+1}, and so on, are in A.

Theorem 2

In a Hausdorff space, a sequence has at most one limit.

The proofs of these two theorems are left as an exercise. There are many other stronger separation properties, which we lack space to treat properly. One at least demands mention.

Definition

A topological space is **normal**, *when it is* T_1, *and given any two disjoint closed sets* F_1 *and* F_2, *there are disjoint open sets* A_1 *and* A_2 *such that* $F_1 \subseteq A_1$ *and* $F_2 \subseteq A_2$.

Normal spaces are important because they have many continuous functions. The following theorem is evidence of this. It is one of several theorems that for reasons of space are given without proof in this chapter.

Urysohn's Lemma

Let F_1 *and* F_2 *be disjoint closed sets in a normal topological space* $\mathscr{S}$. *Then there is a continuous real-valued function* f *on* $\mathscr{S}$ *such that* f *is equal to zero on* F_1 *and equal to one on* F_2.

Exercises

1. Verify the statements made in this section concerning the separation properties of the indiscrete, sector, and modified either/or topologies. What separation properties, if any, does the discrete topology have?
2. Which of the examples of topologies mentioned in the previous exercise are normal? Which are metric?
3. Prove Theorems 1 and 2.
4. Prove that metric spaces are normal.
5. Prove that the product of T_i spaces is T_i, where i is 0, 1, or 2.

COUNTABILITY PROPERTIES

A set is called **countable** when its elements can be arranged in a sequence. Thus the properties to be introduced here can also be called sequential properties. Sequences have played a major role in all the topological proofs in this book, but they are not equally important in all topological spaces. Their usefulness depends on the satisfaction of some form of countability axiom by the space.

Definition

Let $\mathscr{S}$ be a topological space with neighborhood system **N**. $\mathscr{S}$ *is called* **first countable** *if* **N** *can be chosen so that the neighborhoods of each point are countable.* $\mathscr{S}$ *is* **second countable** *if* **N** *itself can be chosen countable.*

As examples of first countable spaces, we have the following theorem.

Theorem B

Metric spaces are first countable.

This is because the neighborhoods $N_{1/k}(P)$, whose radii are $1, \frac{1}{2}, \frac{1}{3}$, and so forth, suffice to define the topology. First countable spaces are precisely the spaces in which sequences are most useful, as the following theorem shows.

Theorem

In a first countable space, a point P is near a set A if and only if P is the limit of a sequence of points from A.

According to this theorem, in a first countable space the nearness relation can be entirely described by limits of sequences. To prove the theorem, suppose that P is near the set A, and let the neighborhoods of P be arranged in a sequence $B_1, B_2, B_3, \ldots$. Since $P \leftarrow A$, $A \cap B_1$ is not empty. Let P_1 be any point in $A \cap B_1$. By (N_2), $B_1 \cap B_2$ contains a neighborhood of P; therefore $A \cap B_1 \cap B_2$ is not empty. Let P_2 be any point in $A \cap B_1 \cap B_2$. Continuing in this way, we choose P_n from $A \cap B_1 \cap \cdots \cap B_n$. The sequence $\{P_n\}$ from the term P_n on is inside the nth neighborhood B_n, therefore $\lim_n P_n = P$. The converse is obvious, so this proves the theorem.

The proof of theorems about countability is greatly eased by the following result.

Lemma (Cantor)

The union of a sequence of countable sets is countable.

Let A_n be a sequence of countable sets, and let the elements of A_n be arranged in a sequence $A_n = \{a_{n,m}\}$. Then the union $A = \cup_n A_n$ can be arranged in a sequence as follows: $A = \{a_{1,1}, a_{1,2}, a_{2,1}, a_{3,1}, a_{2,2}, a_{1,3}, \ldots\}$. In other words, we first list all a's whose subscripts sum to 2, then all whose subscripts sum to 3, and so on following the diagonal scheme shown below:

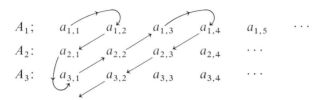

This proof is called **Cantor's first diagonal method**. It follows from this lemma that the set Q of all rational numbers among others is countable, since Q can be considered the union of a sequence of countable sets $Q = \cup_n Q_n$, where Q_n is the set of all rationals with denominator equal to n, $Q_n = \{1/n, 2/n, \ldots\}$. Lest you believe that all sets are countable, we prove the following theorem.

Theorem (Cantor)

There exists an uncountable set.

Let A be the set of real numbers between zero and one, $A = \{x | 0 \le x < 1\}$. Every such x has a decimal expansion $x = .x_1 x_2 x_3 \ldots$ which, to ensure uniqueness, we assume does not end in repeated 9's. Suppose that A were countable, $A = \{x_1, x_2, x_3, \ldots\}$. Then arrange the sequence of decimal expansions in a square array:

$$x_1 = .x_{1,1} x_{1,2} x_{1,3} \ldots$$

$$x_2 = .x_{2,1} x_{2,2} \ldots$$

A new number y can be constructed $y = .y_1 y_2 y_3 \ldots$ by letting

$$y_k = \begin{cases} 1 & \text{if } x_{k,k} \ne 1 \\ 2 & \text{if } x_{k,k} = 1 \end{cases}$$

The number y is not in the sequence $\{x_n\}$ because it differs from x_n at the nth decimal place. This contradicts the assumption that A is countable, and therefore A is not countable. The method of proof used here is called **Cantor's second diagonal method**. Another countable topological property is given in the following definition.

Definition

A subset A of a space $\mathscr{S}$ is **dense** *if every point of $\mathscr{S}$ is near A, that is, if $\bar{A} = \mathscr{S}$. $\mathscr{S}$ is* **separable** *if it has a countable dense subset.*

The real line is separable; the rationals are a countable dense subset. Separable metric spaces are important because of the following theorem.

Theorem C

For metric spaces, separability is equivalent to second countability.

In the first place, whether metric or not, if a space $\mathscr{S}$ has a countable neighborhood system **N**, then the set formed by choosing one point from each neighborhood is countable and dense (Exercise 8). Conversely, if $\mathscr{S}$ is a separable metric space with a countable dense set $A = \{a_1, a_2, a_3, \dots\}$, then the set **N** of neighborhoods with centers at the points of A and radii equal to 1, $\frac{1}{2}, \frac{1}{3}, \dots$ is a countable neighborhood system (i.e., a base) for $\mathscr{S}$ (Exercise 8).

Exercises

6. Which of the following collections are countable and which are uncountable?

   (a) all topologically distinct surfaces

   (b) all lines in the plane

   (c) all 2×2 matrixes with integer entries

   (d) all matrixes with integer entries

   (e) all finite groups

   (f) all groups

   (g) all subsets of the integers

   (h) all polynomials with integer coefficients

   (i) all power series with integer coefficients

7. Which of the following topologies in the plane are separable, which are first countable, and which are second countable: usual, discrete, indiscrete, sector, and either/or?

8. Complete the proof of Theorem C.

9. Prove the following:

(a) The product of first countable spaces is first countable.

(b) The product of second countable spaces is second countable.

(c) The product of separable spaces is separable.

10. Prove that complexes (§23) are first countable, Hausdorff spaces.

11. Find an example of a nonseparable metric space.

---

§41 COMPACTNESS AGAIN

This section is a continuation of the previous section. The importance of compactness demands separate treatment. Sequential compactness, the rock upon which this book has been based, is inadequate for general application. Its dependence on nearness *for sequences* essentially restricts its usefulness to first countable spaces where the topology can be described by limits of sequences. Roughly speaking, in general topological spaces, sequences just don't converge that often. Thus it is necessary to find an alternative property of more general application.

Definition

*Let $\mathscr{S}$ be a topological space. A collection $\{A_\sigma\}$ of subsets is a **cover** for $\mathscr{S}$ if $\cup_\sigma A_\sigma = \mathscr{S}$. If the subsets A_σ are open, then $\{A_\sigma\}$ is called an **open cover**. The space $\mathscr{S}$ is **compact** if every open cover of $\mathscr{S}$ contains a finite subcover.*

The notion of a cover is natural in topology. We have already encountered several examples of covers: any base **B** is an open cover, including, of course, the cover **O** of all open sets. The set **C** of all closed sets is also a cover. In connection with covers, the following theorem is often useful.

Lindelöf Covering Theorem

In a second countable space, every open cover has a countable subcover.

Let $\{A_\sigma\}$ be a cover for the second countable space $\mathscr{S}$. Let **N** be a countable neighborhood system for $\mathscr{S}$. Each A is a union of neighborhoods. Let $\{B_n\}$ be the sequence of all the neighborhoods needed to form by unions all the sets A_σ. For each B_n choose a set A_n from $\{A_\sigma\}$ such that $B_n \subseteq A_n$. Then $\{A_n\}$ is a countable subcover.

In compact spaces, the conclusion of the Lindelöf covering theorem is strengthened to the existence of a finite subcover. The importance of compactness is that it often permits the replacement of an infinite problem by a finite problem. In order to rephrase the definition of compactness in terms of closed sets, let $\{A_\sigma\}$ be any collection of open subsets of $\mathscr{S}$. Then $\{A_\sigma\}$ is a cover for $\mathscr{S}$ if and only if the collection of closed sets $\{A_\sigma'\}$ has empty intersection. This observation leads to the following alternative definition.

Definition

*A collection of closed subsets $\{F_\sigma\}$ of a topological space $\mathscr{S}$ has the **finite intersection property** if every finite intersection of sets from $\{F_\sigma\}$ is not empty. The space $\mathscr{S}$ is **compact** if for any collection of closed sets with the finite intersection property there exists a point P common to all the sets F_σ, in other words, the whole collection has nonempty intersection.*

The proof that these two definitions of compactness are equivalent is left as an exercise. This alternative definition can be applied in a manner similar to the way sequential compactness has been used throughout this book. The following result, which has been used several times already, is typical.

Nested Interval Theorem

Let $\{I_n\}$ be a sequence of nonempty closed intervals on the real line. If $I_{n+1} \subseteq I_n$, then there exists a point x inside all the intervals.

The sequence of intervals obviously has the finite intersection property. Thus this theorem follows straight from the compactness of the closed intervals. Actually we have not proved yet that a closed interval is compact, only that it is sequentially compact. This gap will be closed in the next section.

The following theorems present the two most important properties of compactness.

Theorem

If $\mathscr{S}$ is a compact space and $f:\mathscr{S} \to \mathscr{T}$ is a continuous transformation from $\mathscr{S}$ onto $\mathscr{T}$, then $\mathscr{T}$ is also compact.

Let $\{A_\sigma\}$ be an open cover of $\mathscr{T}$. Then $\{f^{-1}(A_\sigma)\}$ is an open cover of $\mathscr{S}$ from which we extract a finite subcover $\{f^{-1}(A_n)\}$. Then $\{A_n\}$ is a finite subcover of $\mathscr{T}$.

Tychonoff Product Theorem (finite form)

If $\mathscr{S}$ and $\mathscr{T}$ are compact, then the product space $\mathscr{S} \times \mathscr{T}$ is also compact.

The infinite product version of this theorem is one of the most famous theorems of point set topology. Consider first a covering of $\mathscr{S} \times \mathscr{T}$ consisting of neighborhoods: $\{A_\sigma \times B_\sigma\}$. For each point $P \in \mathscr{S}$, the subset $\{P\} \times \mathscr{T}$ of $\mathscr{S} \times \mathscr{T}$ is topologically equivalent to $\mathscr{T}$ and hence compact. Therefore a finite number of sets, say $\{A_n \times B_n\}$, suffice to cover $\{P\} \times \mathscr{T}$. The intersection $A_P = \cap_n A_n$ is an open set containing P such that actually the whole set $A_P \times \mathscr{T}$ is covered by $\{A_n \times B_n\}$. The collection of sets $\{A_P\}$ is a covering of $\mathscr{S}$; therefore there exists a finite set of points P such that the corresponding sets A_P cover $\mathscr{S}$. The union of the finite collections $\{A_n \times B_n\}$ corresponding to these finite number of points P from $\mathscr{S}$ gives a finite cover of $\mathscr{S} \times \mathscr{T}$.

Now consider an arbitrary covering $\{W_\sigma\}$ of $\mathscr{S} \times \mathscr{T}$. For each point (P, Q) of $\mathscr{S} \times \mathscr{T}$, choose a neighborhood of the form $A \times B$ contained in one of the sets W_σ. The totality of neighborhoods thus assembled covers $\mathscr{S} \times \mathscr{T}$, so by the preceding paragraph a finite subcover can be extracted from these neighborhoods. The corresponding containing sets W_σ form a finite subcover of the given cover.

Such is the importance of compactness that considerable attention has been given to the problem of **compactification**, that is, the process of adding points to a given topological space $\mathscr{S}$ in order to make it compact. The simplest compactification arises by adding just one point, usually denoted by the symbol ∞. The neighborhoods of ∞ are defined to be the complements of the compact subsets of $\mathscr{S}$. The resulting space, which we will call $\mathscr{S}^*$, is called the **one-point compactification** of $\mathscr{S}$. It is a generalization of the process by which a point is added to a plane to make a sphere.

Exercises

1. Let **K** denote the collection of compact subsets of a topological space $\mathscr{S}$. Show that **K** satisfies all but one of the closed set axioms (C_1), (C_2), and (C_3).

2. Show that every closed subset of a compact space is compact.

3. Find an example of a subset of the plane (in an unusual topology) that is compact but *not* closed.

4. Prove that the product of sequentially compact spaces is sequentially compact.

5. Prove that $\mathscr{S}^*$ is a compact topological space.

6. If $\mathscr{S}$ is a compact space, show that $\mathscr{S}^*$ is disconnected. On the other hand, if $\mathscr{S}$ has *no* compact sets (other than finite subsets), show that ∞ is near every infinite subset of $\mathscr{S}$.

COMPACT HAUSDORFF SPACES

These include many of the spaces commonly arising in applications, for example, the compact surfaces. Many familiar properties of compact sets appear first in connection with Hausdorff spaces. The following theorem is an example.

Theorem

Compact subsets of Hausdorff spaces are closed.

Let $\mathscr{S}$ be a Hausdorff space with a compact subset K. Let $P \in K, Q \in K' = \mathscr{S} - K$. Then there exist neighborhoods A and B such that $P \in A$, $Q \in B$, and $A \cap B = \varnothing$. Holding the point Q fixed, the collection of sets A, as P ranges over all points of K, covers K. Let $\{A_n\}$ be a finite subcover, and let $\{B_n\}$ be the corresponding neighborhoods of Q such that $A_n \cap B_n = \varnothing$. Let $A^+ = \cup_n A_n$ and $B^+ = \cap_n B_n$. Then A^+ and B^+ are open sets such that $K \subseteq A^+$, $Q \in B^+$, and $A^+ \cap B^+ = \varnothing$. Since Q was any point of K', this proves that K' is open; therefore K is closed.

Corollary

A compact Hausdorff space is normal.

Let $\mathscr{S}$ be a compact Hausdorff space, and let F and K be disjoint closed subsets. Both F and K are actually compact by Exercise 2. Let Q be a point of F. According to the proof of the preceding theorem, there exist open sets A^+ and B^+ such that $K \subseteq A^+, Q \in B^+$, and $A^+ \cap B^+ = \varnothing$. The collection of sets B^+ as Q ranges over all points of F clearly covers F. Let $\{B_n{}^+\}$ be a finite subcover, and let $\{A_n{}^+\}$ be the corresponding "neighborhoods" of K. Let $A^{++} = \cap_n A_n{}^+$ and $B^{++} = \cup_n B_n{}^+$. Then A^{++} and B^{++} are open sets such that $K \subseteq A^{++}$, $F \subseteq B^{++}$, and $A^{++} \cap B^{++} = \varnothing$.

Just how close compact Hausdorff spaces come to the tranditional spaces of analysis is indicated by the following theorem, whose proof is omitted.

Metrization Theorem (Urysohn)

A compact Hausdorff space that is second countable is a metric space.

Exercises

7. Show that a subset K of a compact Hausdorff space is closed if and only if K is compact.

8. Let $f : \mathscr{S} \to \mathscr{T}$ be a continuous transformation, and suppose that $\mathscr{T}$ is a Hausdorff space. Prove that if K is compact in $\mathscr{S}$, then $f(K)$ is closed in $\mathscr{T}$.

9. Let $\mathscr{S}$ be a compact space, $\mathscr{T}$ a Hausdorff space. Prove that a continuous transformation $f : \mathscr{S} \to \mathscr{T}$ is **closed**, meaning that if F is a closed subset of $\mathscr{S}$, then $f(F)$ is a closed subset of $\mathscr{T}$. Show that if f is one-to-one, then f is a topological transformation; hence both $\mathscr{S}$ and $\mathscr{T}$ are compact Hausdorff spaces.

10. Just as familiar properties of compact sets emerge in Hausdorff spaces, so familiar properties of sequentially compact sets emerge in first countable spaces. Prove in a first countable space $\mathscr{S}$ that sequentially compact subsets are closed.

11. Let f be a continuous transformation from a sequentially compact space $\mathscr{S}$ into a first countable space $\mathscr{T}$. Prove that f is closed. Conclude that if f is one-to-one, then both $\mathscr{S}$ and $\mathscr{T}$ are sequentially compact, first countable spaces.

LOCAL PROPERTIES

If every point P of a topological space $\mathscr{S}$ has a neighborhood with a given property, then $\mathscr{S}$ is said to possess that property *locally*. For example, $\mathscr{S}$ is called **locally connected** if every point of $\mathscr{S}$ has a connected neighborhood. A space is called **locally Euclidean** if every point has a neighborhood equivalent to a neighborhood in a Euclidean space R^n of some dimension. The surfaces are locally Euclidean of dimension two. A space is **locally compact** if every point has a neighborhood whose *closure* is compact.

Exercises

12. Prove that the product of locally compact (or locally Hausdorff or locally connected) spaces has the same property.

13. Let $\mathscr{S}$ be a Hausdorff space. Prove that the one-point compactification $\mathscr{S}^*$ is Hausdorff if and only if $\mathscr{S}$ is locally compact.

§42 COMPACT METRIC SPACES

These are the spaces of major interest in analysis. As a metric space is Hausdorff (Theorem A), these spaces have all the properties of compact Hausdorff spaces. In order to investigate other properties, let $\mathscr{S}$ be a compact metric space and $\varepsilon > 0$ any positive number. Consider the covering of $\mathscr{S}$, $\{N_\varepsilon(P)\}$, consisting of all neighborhoods of radius ε. Extracting a finite subcover $\{N_\varepsilon(P_n)\}$, we obtain a finite set $\mathscr{D}_\varepsilon = \{P_n\}$ such that every point of $\mathscr{S}$ is within a distance ε of some point of $\mathscr{D}_\varepsilon$. The set $\mathscr{D}_\varepsilon$ is called **ε-dense** in $\mathscr{S}$. The union $\cup_n \mathscr{D}_{1/n}$ is actually dense in $\mathscr{S}$, and also countable; therefore $\mathscr{S}$ is separable. Thus in addition to being first countable (Theorem B), a compact metric space is second countable (Theorem C). Furthermore, we have the following result.

Theorem

A compact metric space is sequentially compact.

Let $\{Q_m\}$ be any sequence in $\mathscr{S}$. Let $\{N_\varepsilon(P_n)\}$ be the finite cover constructed above. One of the sets of this cover, say $N_\varepsilon(P)$, must contain an infinite number of points from the sequence $\{Q_m\}$, or in other words, a whole subsequence $\{Q_{1,m}\}$. Let $F_1 = \overline{N_\varepsilon(P)}$. Then F_1 is a compact metric space. Let $\mathscr{D}_{\varepsilon/2}$ be an $\varepsilon/2$-dense subset of F_1. One of the sets $N_{\varepsilon/2}(P)$ must contain a subsequence $\{Q_{2,m}\}$ of $\{Q_{1,m}\}$. Letting F_2 be the closure of this neighborhood and continuing in this fashion, we obtain a sequence of closed sets $F_1 \supseteq F_2 \supseteq F_3 \supseteq \ldots$, where F_{n+1} is the closure of a neighborhood of radius $\varepsilon/2^n$, and a sequence of sequences $\{Q_{1,m}\} \supseteq \{Q_{2,m}\} \supseteq \{Q_{3,m}\} \supseteq \cdots$, where $\{Q_{n,m}\} \subseteq F_n$. It is easy to show using compactness that the intersection $\cap_n F_n$ contains at least one point P that is near the original sequence $\{Q_m\}$. Therefore $\mathscr{S}$ is sequentially compact.

Conversely, we have the following theorem.

Theorem

A sequentially compact metric space is compact.

We first prove that $\mathscr{S}$ has a finite ε-dense set $\mathscr{D}_\varepsilon$. If not, then there would be an infinite sequence $\{Q_m\}$ whose points remain a distance ε from each other (Exercise 2). Such a sequence cannot have any near points, contradicting the fact that $\mathscr{S}$ is sequentially compact. Therefore $\mathscr{S}$ has a finite ε-dense set $\mathscr{D}_\varepsilon$, and it follows as before that $\mathscr{S}$ is separable.

Let $\{A_\sigma\}$ be an open covering of $\mathscr{S}$. Since $\mathscr{S}$ is separable, it is also second countable, so that by the Lindelöf covering theorem, there exists a countable subcover $\{A_n\}$. Suppose that $\{A_n\}$ has *no* finite subcover. Choose a point Q_m in $\mathscr{S} - (A_1 \cup A_2 \cup \cdots \cup A_m)$. Let P be a point near the sequence $\{Q_m\}$. Since $\{A_n\}$ is a cover, P belongs to one of the sets A_n. Then A_n must contain an infinite number of terms of the sequence $\{Q_m\}$, which is impossible. Therefore $\{A_n\}$ does have a finite subcover, and $\mathscr{S}$ is compact.

Corollary

For metric spaces, compactness and sequential compactness are equivalent.

Therefore all the spaces discussed in earlier chapters and proved sequentially compact are also compact.

Exercises

1. Let $F_1 \supseteq F_2 \supseteq F_3 \supseteq \cdots$ be a nested sequence of closed sets in a compact metric space. Prove that $\cap_n F_n$ is not empty. Suppose in addition that F_n is the closure of a neighborhood of radius ε_n and that $\lim_n \varepsilon_n = 0$. Prove that $\cap_n F_n$ contains just one point.

2. Suppose the metric space $\mathscr{S}$ has no finite ε-dense set. Construct a sequence $\{Q_n\}$ such that $d(Q_n, Q_m) > \varepsilon$ if $n \neq m$.

3. A sequence $\{Q_n\}$ in a metric space is called a **Cauchy sequence** if given any $\varepsilon > 0$, there is an integer N such that $d(Q_n, Q_m) < \varepsilon$ for n and m greater than N. The space $\mathscr{S}$ is called **complete** if every Cauchy sequence has a limit in $\mathscr{S}$. Prove that a compact metric space is complete.

The following interesting theorem, which we state without proof, shows how the topological properties introduced in this book can be combined to give a characterization of a familiar space.

Bing's Theorem

If the topological space $\mathscr{S}$ has the following properties:

(a) $\mathscr{S}$ is compact

(b) $\mathscr{S}$ is a metric space

(c) $\mathscr{S}$ is connected

(d) $\mathscr{S}$ is locally connected

(e) if P and Q are any two points of $\mathscr{S}$, then $\mathscr{S} - \{P, Q\}$ is still connected

(f) if $\mathscr{J}$ is a Jordan curve in $\mathscr{S}$, then $\mathscr{S} - \mathscr{J}$ is disconnected

then $\mathscr{S}$ is topologically equivalent to a sphere.

Exercise

4. Prove the converse of Bing's theorem by showing that a sphere actually has all the properties listed above.

Notes. This supplement owes much to the survey of point set topology contained in the first chapter of Lefschetz's book [21]. You will probably find the book by Kelley [16] more accessible. It contains a detailed treatment of all the topics discussed here and much more.

Hints and Answers
for Selected
Problems

§1

4. If no cell can be sewn to itself, an annulus requires two cells (no matter how many holes) and a torus requires two cells. If this rule is disregarded, one cell is sufficient for all these figures.

5. An n-holed annulus requires $n + 1$ cells (arrange the holes in a circle around the extra cell). A torus requires four cells.

11. When $a = 2$, an example of this degenerate polyhedron is given by the pattern formed by the sectors under the skin of an orange. When $b = 2$, an example is the polyhedron formed by northern and southern hemispheres when the equator is divided into n edges.

12. Consider the holes as faces that have been removed.

§2

1. (a) A. (b) $[0, 1]$. (c) C. (d) C. (e) the whole plane. (f) F. (g) G. (h) H plus the origin. (i) J.

2. The sets A, B, C, D, E, F, H, J are thick. G and J are thin.

3. Use a sequence of neighborhoods of P with smaller and smaller radii.

8. Given a point P, define the set $A = \{Q| \, \|f(Q) - f(P)\| \geq \varepsilon\}$ and show that $f(P)$ is not near $f(A)$ in order to conclude that P is not near A. Then an application of Exercise 6 should lead to the desired result.

9. Let P be a point of D and let A be a subset of D such that $P \leftarrow A$. Supposing, contrary to what we want to prove, that $f(P)$ is not near $f(A)$, use Exercise 6 to find an $\varepsilon > 0$ such that no point within the distance ε of $f(P)$ is in $f(A)$. Then use the $\varepsilon - \delta$ definition of continuity to deduce the contradiction that P is not near A.

11. Only the stretching transformations, including the two infinite stretches, are topological. The others are topological on parts of their domains.

12. (a) $t(P) = P + (a, b)$.   (b) Let h be the translation by one unit to the right, $h(P) = P + (1, 0)$; let j be a stretching by a factor of $\pi/2$ in the y direction, $j(P) = (x, \pi y/2)$; and let t be the transformation of Example 3, $t(x, y) = (x \cos(y), x \sin(y))$. Then the desired transformation is the composition $k = t \circ j \circ h$,

$$k(P) = \left((x + 1) \cos\left(\frac{\pi y}{2}\right), (x + 1) \sin\left(\frac{\pi y}{2}\right) \right)$$

(c) Use a stretch by π instead of by $\pi/2$.   (d) Stretch, translate, and rotate.

§3

1. Let P be any point of S. P is in some square R_N. Let d be the distance from P to the closest of the finite number of points $P_1, P_2, \ldots, P_N$. Any neighborhood of P that is contained in R_n and has a radius less than d cannot contain any point of the sequence $\{P_n\}$ (other than P itself perhaps). This proves that the sequence has no near points.

2. Let N_1 be a neighborhood of P. Since $P \leftarrow S$, N_1 contains a point P_1 of S. Let N_2 be a neighborhood of P not containing P_1 (draw a picture). Let P_2 be a point of N_2 from S. Let N_3 be a neighborhood of P not containing P_1 or P_2. Let P_3 be a point of S in N_3. The sequence $\{P_n\}$ constructed in this way has the desired property.

3. 3 are compact, 2 are only bounded, 3 are only closed, and one is nothing.

4. Consider an infinite stretch.

6. Show that if a point P is near the closure of S, then P is also near S itself.

8. Show that if a connected subset of the real line contains the points P and Q, then it must also contain all the points between P and Q.

§4

The exercises involving the finite, sector, and either/or topologies are made easier if one first notices (and proves) the following facts: (a) In the finite topology, every point is near an infinite set but finite sets have no near points.   (b) In the sector topology, every point is near any set that contains a point directly above it.   (c) In the either/or topology, the points on the x-axis are near every set that contains a point not on the x-axis.

9. usual: TTTT; discrete: TTTT; indiscrete: FFTF; finite: TFFF; sector: FFFF; either/or: FFFT.

§5

2. (a) The transformation is basically a swirling motion away from the x-axis. Solutions of the corresponding system of differential equations are $x = t$, $y = Ke^t$; thus the solution curves are the exponential curves $y = Ke^x$. (c) The solutions are $x = Ke^t$, $y = He^{-t}$; thus the solution curves are the hyperbolas $xy = k$. (f) The system of differential equations $x' = y$, $y' = x$ leads to the second order equation $x'' = x$ with the solutions $x = ae^t + be^{-t}$, $y = ae^t - be^{-t}$. The solution paths are the hyperbolas $x^2 - y^2 = K$.

§6

2. the annulus

4. Because f transforms D into itself, all the vectors V are inside the triangle.

6. The two statements of the topological lemma are contrapositives of each other.

§7

1. (a) unstable node at the origin, the integral paths are straight lines. (b) center at the origin, integral paths are circles. (c) still a center, integral paths are ellipses. (f) two nodes at $(\pm 1, 0)$. (g) two saddle points at $(\pm 1, 0)$. (h) node $(1, 0)$ and saddle point $(-1, 0)$. (i) center $(1, 0)$ and saddle point $(-1, 0)$. (j) This differential equation is homogeneous; the critical point is a dipole (see Figure 9.2). (k) This is a first-order linear differential equation; the critical point is a saddle point.

4. Use the index lemma.

§8

1. (a) 1. (b) -1. (c) 0. (d) 0.

2. (a) 1. (b) 0. (c) -1. (d) 3.

§9

2. (a) 3. (b) 1. (c) 2. (d) -2. (e) 0. (f) 0.

6. What goes in must go out.

8. Work a few examples of the different types of sector in order to get the idea. Note that the vector always turns just one direction within each sector.

9. Use the original definition of winding number.

11. There is only one of index 3, 5 of index 2, and 8 of index 1. Be careful: two critical points that are reflections of each other are topologically equivalent.

§10

3. The confluence of a saddle point and a center is a critical point of index 0, in this case having two hyperbolic sectors.

§11

1. On the leg BC, $U(x, L)$ points in the direction of the secant to P from the point X. As x varies from L to 0, X travels once clockwise around γ, and so the vector $U(x, L)$ makes a half revolution clockwise. On the leg CA, $U(0, y)$ points in the direction of the secant from P to a point Y. As y varies from L to 0, Y travels once clockwise around γ, and so the vector $U(0, y)$ makes a second half revolution clockwise.

3. (a) $\phi = (x^2 + y^2)/2$.   (d) $\phi = x^3/3 + x - xy^2$.   (f) $\phi = x^3 y - xy^3$.

4. Euler's formula for polyhedra.

5. Don't forget the saddle points where the streams run out of the lakes!

§12

3. It is not possible to predict the numbers of sectors of each type for V^* given these numbers for V: consider examples of critical points with two each of elliptic and hyperbolic sectors with different arrangements. There is also no connection between stability for V and for V^*: centers can be dual to stable or unstable nodes.

7. Since V cannot have nodes or elliptic sectors, V^* cannot have centers or elliptic sectors.

§13

1. Consider the special case when γ is the unit interval. Then use the fact that connectedness is a topological property.

3. If a rectangle were filled by a path, then removing one point would produce a disconnected set.

4. Use the ideas behind the proof of the polygonal chain theorem.

§14

3. A 2-chain that contains the unbounded face is not compact.

4. (a) $C = C_1 + C_2$.   (b) $C = C_1 + C_5 = C_2 + C_3$.   (c) $C = C_4 + C_5 = C_1$.

§15

2. Use additivity.

3. For 2-chains, verify this equation for simplexes first and then use additivity.

11. Use Exercise 3.

§16

3. This does not contradict the fundamental lemma because $S_1 + T_1 = S_2 + T_2$ and $S_1 + T_2 = S_2 + T_1$.

6. Let S_1 be the inside 2-chain for λ_1, let S_2 be inside for λ_2, and let S_3 be inside for λ_3. Let T_1, T_2, and T_3 be the corresponding outside 2-chains. Then $\lambda_1 + \lambda_2 + \lambda_3$ is the boundary of $T_1 + T_2 + T_3$ (in G), so that $\lambda_3 \sim \lambda_1 + \lambda_2$ (in G).

7. Using the same notation as the preceeding answer, if λ is any 1-cycle on $\mathcal{G}$, let T be its outside 2-chain. Among the eight 2-chains $\varnothing$, the whole plane, S_1, S_2, S_3, T_1, T_2, and T_3, exactly one contains the same holes as T. Depending on which one this is, λ is homologous to $\lambda_1 + \lambda_2 + \lambda_3$, $\varnothing$, $\lambda_2 + \lambda_3$, $\lambda_1 + \lambda_3$, $\lambda_1 + \lambda_2$, λ_1, λ_2, or λ_3, respectively.

9. There are 2^n cycles in a representing set and n cycles in a homology basis.

§17

2. Let the cell correspond to a unit square. Imitate the proof of the Jordan curve theorem for paths, bisecting the square alternately vertically and horizontally.

§18

1. Use the Jordan curve theorem and the result of Exercise 2 of §17.

§19

1. (a) a disk.   (b) a cylinder.

6. Place the hole in the center of the plane model of the projective plane. Two cuts are necessary connecting the hole with the two vertexes of the plane model.

8. No matter how the sides of the rectangle are identified, the result is equivalent to a sphere, projective plane, torus, or Klein bottle. If some sides are left unidentified, then the possibilities are a disk, a cylinder, or a Möbius strip. With a triangle, one side must be left unidentified, so the only possibilities are a disk and a Möbius strip.

9. Only (b) is not regular. The others have 3-, 4-, and 6-gons, respectively, for faces and 6, 4, and 3 edges, respectively, meeting at each vertex.

11. (a) The corner is a node, while the two critical points on the edges are saddle points.   (b) There are four saddle points in addition to the two dipoles.   (c) The two critical points on the edges are saddle points (note that one of them has two parabolic sectors); the corner is a node.

12. Does not, does, does not, does, does not.

13. The $2n$ cuts can be the edges of the plane model.

§20

1. (a) There are two triangles PQR.   (b) There are four edges PQ (there should be only two, since the projective plane is a surface).

3. The hole cut out from each triangulable space can be a triangle.

5. (b) and (e) are not surfaces.

8. (d) is not connected.

§21

4. Figure 21.1 is a projective plane. Notice that the five edges $c:d:e:f:b$ can be treated as a single unit. The other surfaces are as follows: (a) sphere, (c) projective plane, (d) two spheres, (f) sphere, (g) torus, (h) sphere, (i) the connected sum of two projective planes, (j) the connected sum of three projective planes.

6. A pair of edges changes type when a cut separates the two edges and one piece is turned over before regluing. This cannot affect the proof of the classification theorem because once a pair is normalized, it is never separated by a cut.

9. According to the classification theorem, it will suffice to show that a sphere, a connected sum of tori, and a connected sum of projective planes can be put in this form. A sphere is already in the desired form. So is the torus. That a connected sum of tori can also be placed in this form can now be proved by induction. The induction step consists of taking the connected sum of a torus and the connected sum of $(n-1)$ tori already in the desired form. After one cutting and pasting operation, this will also be in the desired form. The connected sum of projective planes can be handled using the preceding exercise.

§22

1. (a) Möbius strip. (b) disk. (c) cylinder.

§23

1. As a check, here are the numbers of vertexes in each complex: (a) 3, (b) 2,
(c) 1, (d) 10, (e) 10. The mystery hexagon is the connected sum of three projective
planes.

2. The outside can be constructed from two octagons identified at every other edge.
The vertical cylinder inside can be constructed from two rectangles, and the hole in the
hole can be constructed from four hexagons (in other words the hole in the hole is cut
in half twice: once through the equator and again across the torus in the middle).
Incidentally, this surface is topologically equivalent to the connected sum of three tori.

5. A pedestrian on a walk arriving at a vertex of even degree can always find an
untrodden edge by which to leave, no matter how often he or she has already visited
that vertex.

12. (a) Edges are identified in groups of four, vertexes in groups of six. This is a
manifold. (b) Edges are identified in groups of four and all vertexes are identified
together. This is not a manifold.

§24

1. For $\mathscr{C}_3$, a and b are generators. For $\mathscr{C}_4$, a and c are generators. For $\mathscr{C}_5$, a, b, c,
and d are generators. For $\mathscr{C}_6$, a and e are generators.

2. (a) This figure has only one symmetry operation: a reflection. Its symmetry
group is isomorphic to $\mathscr{C}_2$. (b) This figure has the same symmetry as Figure 24.3.
(c) This figure is invariant under rotations of 120° and 240°. The symmetry group is
isomorphic to $\mathscr{C}_3$. For the remaining parts of this exercise, here are the small groups
isomorphic to the symmetry groups: (d) $\mathscr{C}_2$. (e) $\mathscr{C}_1$. (f) $\mathscr{C}_5$. (g) $\mathscr{D}_3$. (h) $\mathscr{C}_2$.
(i) $\mathscr{D}_2$. (j) $\mathscr{C}_6$. (k) $\mathscr{C}_2$. (l) $\mathscr{C}_1$. (m) $\mathscr{C}_4$. (n) $\mathscr{D}_2$. (o) $\mathscr{C}_5$. (p) $\mathscr{C}_2$. (q) $\mathscr{C}_4$. (r) $\mathscr{C}_5$.
(s) $\mathscr{D}_3$. (t) $\mathscr{C}_2$.

6. $\mathscr{C}_2, \mathscr{D}_2$.

8. (a) $A + B$ is one of many 2-chains not a 2-cycle. (b) $A + B + C + D + E + F$
is actually the only 2-cycle. It is not a 2-boundary. (c) $a + b$. (d) $a + b + c + d$.
(e) The only 2-chains are $\varnothing$ and A, both of which are 2-cycles. (f) The only 1-chains
are $\varnothing$, a, b, and $a + b$. All are 1-cycles. (g) This is simply a reflection of the fact that
all 1-chains are 1-cycles. (h) $a + b$ is a 1-cycle that is not a 1-boundary; the only
1-boundary is $\varnothing$. (i) yes.

9. Consider 0-chains with only one and two vertexes at first. Remember that $\mathscr{K}$ is
connected.

12. From a homological point of view these complexes have just one vertex, since in $\mathcal{N}$, $\mathcal{O}$, and $\mathcal{P}$, $P \sim Q$. Therefore in all cases $H_0 \cong \mathscr{C}_2$. For $\mathcal{N}$, $\mathcal{O}$, and $\mathcal{P}$, the only 2-cycle is $\varnothing$; therefore in these cases $H_2 \cong \mathscr{C}_1$. In contrast, $H_2(\mathcal{M}) \cong \mathscr{C}_2$. The interesting group is H_1. For $\mathcal{M}$, a, b, and $a + b$ are cycles, so like the torus, $H_1(\mathcal{M}) \cong \mathscr{D}_2$. For $\mathcal{N}$, b is a cycle but a is not. However, b is also a boundary, and hence $b \sim \varnothing$. Thus $H_1(\mathcal{N}) \cong \mathscr{C}_1$. For $\mathcal{O}$, no edge by itself is a cycle. The cycles are $a + b$, $b + c$, and $a + c$. However, $b + c \sim \varnothing$, since $b + c$ is the boundary of the Möbius strip. Therefore $H_1(\mathcal{O}) \cong \mathscr{C}_1$. Finally for $\mathcal{P}$, b, c, and $b + c$ are all 1-cycles. However, $b + c \sim \varnothing$, or $b \sim c$; thus homologically speaking there is only one nonzero cycle and $H_1(\mathcal{P}) \cong \mathscr{C}_2$.

13. In all cases $H_0 \cong \mathscr{C}_2$. K_3 has just one 1-cycle, so $H_1(K_3) \cong \mathscr{C}_2$. $\mathcal{R}$ has three 1-cycles, $H_1(\mathcal{R}) \cong \mathscr{D}_2$. $\mathcal{T}$ has no 1-cycles, so $H_1(\mathcal{T}) \cong \mathscr{C}_1$.

§25

1. The pasting operation in Step 2 of the classification theorem closes off the cycle of triangles around a vertex. Therefore the edge removed in this operation cannot be included in a 1-cycle and hence cannot affect the homology of the surface.

2. The fact that all vertexes are identified to one implies $H_0(\mathscr{S}) \cong \mathscr{C}_2$. The fact that the polygon has edges identified in pairs means that the polygon itself forms a 2-cycle. Therefore $H_2(\mathscr{S}) = \mathscr{C}_2$.

§26

2. If a linear combination of elements of A equals zero and the coefficient of one element is one, then that element can be transferred to the other side of the equation; hence it is dependent on the other elements of A.

3. If an element x is equal to two different linear combinations of elements from B, then by subtracting them we obtain a linear combination of elements of B that equals zero.

8. Use the normal forms. The h_1 curves may be taken as the h_1 boundary edges of the normal form. When the surface is cut along these paths, it is not disconnected but becomes a disk.

9. Each boundary curve is a 1-cycle, but the set of boundary cycles is not independent but rather being equal to the boundary of the whole surface, the sum of the boundary cycles is homologous to zero. Therefore any one boundary curve may be regarded as being dependent on the others, which are independent. Thus if b is the number of boundary curves, then the Betti number of the surface with boundary is increased by $(b - 1)$ over the first Betti number of the corresponding surface without boundary.

11. For the n-page book: $h_0 = 1$, $h_1 = n$ (the set of boundary curves of the pages forms a set of n independent 1-cycles), and $h_2 = 0$. For the box with n compartments: $h_0 = 1$, $h_1 = 0$ (even without the compartment dividers, every 1-cycle is a boundary already on the box frame, which is topologically equivalent to a sphere), and $h_2 = n$ (the compartments form a set of n independent 2-cycles).

16.

Table of Euler characteristics	Without boundary	With b boundary curves
Sphere	2	$2 - b$
Connected sum of n tori	$2 - 2n$	$2 - 2n - b$
Connected sum of n projective planes	$2 - n$	$2 - n - b$

17. For a book with n pages, $\chi = 1 - n$. For a box with n compartments, $\chi = 1 + n$.

18. The graphs need not be connected.

§27

1. Six colors are needed to color all maps on a Möbius strip, the same number as on a projective plane. In general, the same number of colors are needed for surface with boundary as for the corresponding surface without boundary—for the same reason that four colors are needed to color all maps on a disk.

4. One central country surrounded by a ring of five countries.

6. If $2E < 3F$, then $a = 2E/F < 3$; therefore at least one face has only two edges. This is impossible because two vertexes are connected by at most one edge. Therefore $2E \geq 3F$.

7. If n points on $\mathscr{S}$ can be mutually connected by nonintersecting paths, the resulting complex has $V = n$, $2E = n(n - 1)$, and $F \leq 2E/3 = n(n - 1)/3$. It follows that $n \leq N_{\chi}$.

9. Examples of degenerate polyhedra on the projective plane can be found by taking one hemisphere of a degenerate polyhedron and identifying opposite points on the boundary.

10. There are five solutions, the same solutions as for the sphere. Examples of the corresponding complexes can be found by taking one hemisphere of the stereographic projection of the corresponding platonic solid.

§29

10. For the sphere $H_0 = \mathscr{Z}$, $H_1 = \mathscr{C}_1$, $H_2 = \mathscr{Z}$. For the projective plane $H_0 = \mathscr{Z}$, $H_1 = \mathscr{C}_2$, $H_2 = \mathscr{C}_1$. For the double torus $H_0 = \mathscr{Z}$, $H_1 = \mathscr{Z}^4 = \mathscr{Z} \oplus \mathscr{Z} \oplus \mathscr{Z} \oplus \mathscr{Z}$, $H_2 = \mathscr{Z}$. For the connected sum of three projective planes $H_0 = \mathscr{Z}$, $H_1 = \mathscr{Z}^2 \oplus \mathscr{C}_2$, $H_2 = \mathscr{C}_1$. The plane models given in Exercise 9 of §21 give the clearest picture of the homology groups for the twisted surfaces.

§30

7. Let G be a group, and suppose that x and y are torsion elements. Then there exist integers m and n such that $mx = 0$ and $ny = 0$. Let $k = mn$; then $k(x + y) = 0$, proving that $x + y$ is also a torsion element.

9. It is clear that every element in the kernel of f must be a torsion element. Conversely, suppose that x is a torsion element; then by the preceeding exercise $f(x)$ is a torsion element of B. But B is torsion free; therefore $f(x)$ is zero, and x is in the kernel of f.

11.

Table of homology groups of surfaces	Without boundary			With b boundary curves		
	H_0	H_1	H_2	H_0	H_1	H_2
Connected sum of n tori ($n \geq 0$)	$\mathscr{Z}$	$\mathscr{Z}^{2n}$	$\mathscr{Z}$	$\mathscr{Z}$	$\mathscr{Z}^{2n+b-1}$	$\mathscr{C}_1$
Connected sum of n projective planes ($n > 0$)	$\mathscr{Z}$	$\mathscr{Z}^{n-1} \oplus \mathscr{C}_2$	$\mathscr{C}_1$	$\mathscr{Z}$	$\mathscr{Z}^{n+b-1}$	$\mathscr{C}_1$

12. For $\mathscr{C}_3$, let $\mathscr{K}$ be a triangle with all three sides identified. The other groups can be handled in a similar fashion.

15. (b), (c), (f), and (g) are all complete.

§31

3. Note that (c) is a projective plane and therefore not orientable, yet the conclusion of the Poincaré index theorem holds. (d) and (e) contain cross points with 6 and 8 sectors.

7. All orientable surfaces have phase portraits with just one critical point. To draw one, you may find it convenient to use the plane models given in Exercise 9 of §21.

§32

1. The sphere and disk have trivial first homology group; therefore they are their own universal covering space. The first homology group of the cylinder $\mathscr{N}$ is isomorphic to $\mathscr{Z}$. Therefore each of the two vertexes of the plane model gives rise to a doubly infinite sequence of vertexes of the covering space. These sequences of vertexes together with the corresponding sequence of sheets can be laid out as a strip in the plane. This strip is the universal covering space of the cylinder.

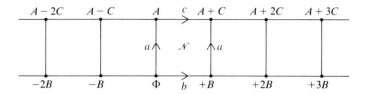

The first homology group of the Klein bottle $\mathscr{K}$ is isomorphic to $\mathscr{Z} \oplus \mathscr{C}_2$. The single vertex of $\mathscr{K}$ gives rise to two sequences of vertexes and sheets. These can be arranged in a double strip. Opposite sides of this strip remain to be identified; therefore the universal covering space of the Klein bottle is an infinite cylinder.

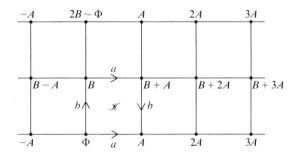

The first homology group of the Möbius strip is isomorphic to $\mathscr{Z}$. Like the cylinder, the covering space is a strip. This is best seen using a triangular model of the Möbius strip $\mathscr{M}$.

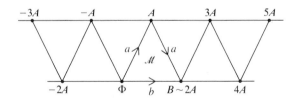

2. Using homology (mod 2), one obtains a four-sheeted covering of the torus. More generally, a covering of any number of sheets can be formed by using a rectangle of sheets from the universal covering space. All of these coverings using just a finite number of sheets are themselves topologically equivalent to tori.

9. (a) translation. (b) rotation by $180°$. (c) reflection across the x-axis. (d) rotation by $90°$. (e) stretch by a factor of 2 in all directions from the origin. (f) reflection across the y-axis. (g) translation. (h) rotation by θ.

10. If $m > n$, set $f(\infty) = 0$; if $m = n$, set $f(\infty) = a_n/b_n$; and if $m < n$, set $f(\infty) = \infty$.

12. Given the point (a, b, c) on the sphere $a^2 + b^2 + c^2 = 1$, the point diametrically opposite is $(-a, -b, -c)$. Under stereographic projection these two points become the points $z = (a + bi)/(1 - c)$ and $z^* = (-a - bi)/(1 + c)$ in the complex plane for which $\bar{z}z^* = (-a^2 - b^2)/(1 - c^2) = -1$.

§33

1. The star of a point in the interior of a triangle is the whole interior of that triangle. The star of a point on an edge is the union of that edge minus its endpoints plus the interior of the triangles containing that edge. The star of a vertex P is the union of all edges containing P minus their other endpoints plus the interiors of all triangles containing P.

§34

1. Without part (b) of the definition of contiguous transformation, the transformation D_1 would not be well defined.

§35

11. The winding number of V on $\mathscr{J}$ equals the one-dimensional Brouwer degree of the transformation f. By the invariance theorem, the degree is independent of the simplicial approximation used to compute $H_1(f)$. The choices (a), (b), and (c) mentioned in this exercise only affect the nature of this simplicial approximation.

§36

2. To prove $C = D$, apply ∂ to equations (1) and use the fact that $\partial \tau_2 = \tau_1 \partial$.

4. Use the identity transformation.

5. The letters o and s are not used. The letters j, m, and h are used two different ways.

6. (a) Fixed points are 0 and ± 1, all of index one.   (c) Fixed points are all the $(n - 1)$ roots of 1, plus zero. All have index one.   (d) The only fixed point is 0 of index two. (f) The only fixed point is 0 of index one.

7. The indexes are   (a) 1,   (b) 1,   (c) 1,   (d) 2.

10. The disk $\mathscr{D}$ has trivial homology groups except that $H_0(\mathscr{D}) = \mathscr{Z}$. Since $H_0(f)$ is the identity homomorphism, it follows that $L(f) = 1$.

12. $\chi(\mathscr{S})$.

14. If f is not onto, $b = 0$.

15. Since the Brouwer degree of $f \circ g$ is the product of the Brouwer degrees of f and g, it is impossible that all three of these degrees equal -1.

17. Apply Exercise 15.

18. The origin is always fixed by f_B. Other fixed points (in some cases) are   (a) $(\frac{1}{2}, \frac{1}{2})$, for a total of two fixed points each of index one;   (c) $(\frac{5}{6}, \frac{1}{2})$, $(\frac{2}{3}, 0)$, $(\frac{1}{2}, \frac{1}{2})$, $(\frac{1}{3}, 0)$, $(\frac{1}{6}, \frac{1}{2})$, for a total of six fixed points;   (e) the whole diagonal of the torus $(x + y = 1)$ is fixed. These are not isolated fixed points, so the Lefschetz fixed point theorem does not apply.

§37

2. $H_k(f)$ maps every element to the identity.

9. Let f be a continuous transformation of the disk, and let V be the corresponding vector field. Let V_t be the vector field $V_t(x, y) = tV(x, y)$. Let p_t be the continuous transformation of the disk whose vector field is V_t. Then p_t is a homotopy of f with the identity transformation.

10. (a) $f_t(x, y) = (e^{-t}x, e^t y)$. The origin is a saddle point.   (b) $f_t(x, y) = (e^t x, y/(1 + xy(1 - e^t)))$. The whole y-axis consists of critical points.

11. (a) centers at $\pm i$.   (b) dipole at 1.   (c) nodes at ± 1.   (d) dipole at ∞.

§38

5. The third axiom is the hard one. Use the ideas in the proof of Lemma 2 (§34).

§39

2. Let A be an open set in $\mathscr{S}$. Every point P of A is not near A' and therefore has a neighborhood B_P entirely contained in A. The union $\cup_P B_P$ is thus equal to A.

4. According to Exercise 2, the open sets are the unions of neighborhoods. If a base $\mathbf{B}$ is used to establish a topology for $\mathscr{S}$, the unions of sets from $\mathbf{B}$ by definition will just be the open sets of the topology previously established on $\mathscr{S}$. Thus this previous topology and the topology established by $\mathbf{B}$ will be the same.

6. (K_1) is not satisfied unless the group is torsion free. (K_3) is not satisfied unless the group is trivial.

7. (I_1) $I(\mathscr{S}) = \mathscr{S}$. (I_2) $I(A) \subseteq A$. (I_3) $I(A \cap B) = I(A) \cap I(B)$. (I_4) $I(I(A)) = I(A)$.

8. The discrete topology is always the strongest, and the indiscrete is the weakest.

9. In the diagram below, each topology is stronger than all the topologies below it with which it can be connected by the given lines.

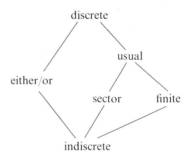

11. $f^{-1}(I(A)) = I(f^{-1}(A))$.

12. Prove and use the fact that $f^{-1}(A \cup B) = f^{-1}(A) \cup f^{-1}(B)$.

13. Draw a picture in the plane as a guide.

15. Use the fact that $d(P, A) = 0$ if and only if there is a sequence of points $\{P_n\}$ from A such that $\lim_n d(P, P_n) = 0$.

19. Use Exercise 12.

20. Draw a picture of the neighborhoods in the plane corresponding to the metric (2).

22. Let P be a point of $\mathscr{S}$. The set $\{P\} \times \mathscr{T}$ in $\mathscr{S} \times \mathscr{T}$ is equivalent to $\mathscr{T}$ and therefore is connected (draw a picture in the plane!). Now use Exercise 10 of §3.

§40

2. Only the discrete topology.

4. Let F_1 and F_2 be disjoint closed sets in a metric space. Let $A_1 = \{P | d(P, F_1) < \frac{1}{3}d(P, F_2)\}$ and $A_2 = \{P | d(P, F_2) < \frac{1}{3}d(P, F_1)\}$. Then A_1 and A_2 are disjoint open sets such that $F_1 \subseteq A_1$ and $F_2 \subseteq A_2$.

6. (b), (f), (g), and (i) are uncountable.

11. The real line with the discrete topology.

§41

2. Let F be a closed subset of a compact space $\mathscr{S}$. Let $\{A_\sigma\}$ be an open cover of F. Then $\{A_\sigma, F'\}$ is an open cover of $\mathscr{S}$. Extracting a finite subcover of $\{A_\sigma, F'\}$, we also obtain a finite subcover of $\{A_\sigma\}$, proving that F is compact.

3. Consider a closed rectangle disjoint from the x-axis in the either/or topology. Since the either/or topology is the same as the usual topology off of the x-axis, this set

is compact. But in the either/or topology, its closure includes the x-axis; therefore it is not closed.

6. If $\mathscr{S}$ is compact, $\{\infty\}$ is a clopen set.

§42

2. Construct the sequence $\{Q_m\}$ term by term. Suppose that $Q_1, Q_2, \ldots, Q_N$ have already been found. Because $\mathscr{S}$ has no finite ε-dense set, there must be a point P which lies at a distance at least ε from all the points $Q_1, Q_2, \ldots, Q_N$. Set $Q_{N+1} = P$.

3. Use Exercise 1.

Suggestions for Further Reading

The notes to individual chapters and the bibliography contain suggestions for those who wish to pursue the topics discussed in the text. Here are listed a few other areas of topology and its applications that may be of interest.

Morse Theory

Homology as applied to maxima and minima. Reference: Milnor, J., *Morse Theory* (Annals of Mathematics Studies, vol. 51, Princeton University Press, Princeton N.J., 1963).

Differential Geometry

A blend of calculus and topology with many important applications to interesting spaces. Reference: Spivak, M., *A Comprehensive Introduction to Differential Geometry* (Publish or Perish, write care of the author, Brandeis University, Waltham, Mass. 02154, 1970.)

Functional Analysis

A blend of analysis and topology and algebra. The spaces involved are algebraic objects such as groups and vector spaces with a topology, too. The goals of the theory, which has become a major branch of mathematics, include the solution of differential equations. Reference: Simmons, G., *Introduction to Topology and Modern Analysis* (McGraw-Hill, New York, 1963). Also N. Dunford and J. Schwartz, *Linear Operators* (vol. I, Inter-science, New York, 1957).

Homological Algebra

A branch of abstract algebra based on the algebraic ideas involved in homology. Reference: D.G. Northcott, *A First Course of Homological Algebra* (Cambridge University Press, New York, 1973). Also MacLane, S., *Homology* (Springer-Verlag, New York, 1975).

Bibliography

The list given here is restricted to works actually used in preparing this book. Readers desiring to investigate further the qualitative theory of differential equations or homology theory should consult the bibliographies in [13] and [29], respectively.

[1] Appel, K., and W. Haken. "Every Planar Map Is Four Colorable." *Bull. A.M.S.* 82 (1976), pp. 711–712.

[2] Banchoff, T. F. "Critical Points and Curvature." *Amer. Math. Monthly* 77 (1970), pp. 475–485.

[3] Bergamini, David. *Mathematics.* Time-Life Books (part of the *Life Science Library*), New York, 1963.

[4] Blackett, Donald. *Elementary Topology: A Combinatorial and Algebraic Approach.* Academic Press, New York, 1967.

[5] Cameron, P., J. G. Hocking, and S. A. Naimpally. "Nearness—A Better Approach to Continuity and Limits." *Amer. Math. Monthly* 81 (1974), pp. 739–745.

[6] Courant, Richard, and Herbert Robbins. *What Is Mathematics?* Oxford University Press, London, 1941.

[7] Eilenberg, S., and N. Steenrod. *Foundations of Algebraic Topology.* Princeton University Press, Princeton, N.J., 1952.

[8] Frechet M. and K. Fan. *Initiation to Combinatorial Topology.* Prindle, Weber & Schmidt, Boston, 1967.

[9] Greenberg, M. *Lectures in Algebraic Topology.* W. A. Benjamin, Reading, Mass., 1967.

[10] Harary, F. *Graph Theory.* Addison-Wesley, Reading, Mass., 1969.

[11]   Herstein, I. N. *Topics in Algebra*. Blaisdell, Waltham, Mass., 1964.

[12]   Hilbert, D. and S. Cohn-Vossen. *Geometry and the Imagination*. Chelsea, New York 1952.

[13]   Hirsch, M., and S. Smale. *Differential Equations, Dynamical Systems and Linear Algebra*. Academic Press, New York, 1974.

[14]   Hopf, H. "A New Proof of the Lefschetz formula on Invariant Points." *Proc. Nat. Acad. of Science* 14 (1928), pp. 149–153.

[15]   Kasner, E., and J. Newman. *Mathematics and the Imagination*. Simon & Schuster, New York, 1940.

[16]   Kelley, J. *General Topology*. Van Nostrand, Princeton, N.J., 1955.

[17]   Klein, F. *On Riemann's Theory of Algebraic Functions and Their Integrals*. Dover, New York, n.d.

[18]   Kline, M. *Mathematical Thought from Ancient to Modern Times*. Oxford University Press, Oxford, 1972.

[19]   Lang, S. *Algebra*. Addison-Wesley, Reading, Mass., 1965.

[20]   Lebesgue, H. "Quelques consequences simples de la formule d'Euler." *J. de Math. pures et appl.* 19 (1940), pp. 27–43.

[21]   Lefschetz, S. *Algebraic Topology*. Amer. Math. Soc., New York, 1942.

[22]   Lefschetz, S. *Differential Equations: Geometric Theory*. Interscience, New York, 1957.

[23]   Massey, W. *Algebraic Topology: An Introduction*. Harcourt, Brace and World, New York, 1967.

[24]   Nemytskii, V., and Stepanov. *Qualitative Theory of Differential Equations*. Princeton University Press, Princeton, N.J., 1960.

[25]   Newman, M. H. A. *Elements of the Topology of Plane Sets of Points*, 2nd ed. Cambridge University Press, Cambridge, 1951.

[26]   Sansone, G., and J. Gerretsen. *Lectures on the Theory of Functions of a Complex Variable*. Wolters-Noordhoff, Gronigen, vol. I 1960, vol. II 1969.

[27]   Seifert, H., and W. Threlfall. *Lehrbuch der Topologie*. Chelsea, New York.

[28]   Simmons, G. *Differential Equations with Applications and Historical Notes*. McGraw-Hill, New York, 1972.

[29]   Vick, J. *Homology Theory*. Academic Press, New York, 1973.

Index